# THEORIES of ANIMAL MEMORY

# COMPARATIVE COGNITION AND NEUROSCIENCE

*Thomas G. Bever, David S. Olton, and Herbert L. Roitblat, Series Editors*

# THEORIES of ANIMAL MEMORY

Edited By

## D. F. Kendrick
*Middle Tennessee State University*

## M. E. Rilling
*Michigan State University*

## M. R. Denny
*Michigan State University*

LAWRENCE ERLBAUM ASSOCIATES, PUBLISHERS

1986    Hillsdale, New Jersey                    London

Lawrence Erlbaum Associates, Inc., Publishers
365 Broadway
Hillsdale, New Jersey 07642

**Library of Congress Cataloging-in-Publication Data**
Main entry under title:

Theories of animal memory.

   Bibliography: p.
   Includes index.
   1. Animal memory.   I. Kendrick, D. F. (Donald F.)   II. Rilling, M. E. (Mark E.)
III. Denny, M. Ray (Maurice Ray), 1918–
QL785.2.T48   1986      591.5′1      85-16148

ISBN 0-89859-636-X
ISBN 0-89859-697-1 (pbk.)

Printed in the United States of America

10   9   8   7   6   5   4   3   2   1

# Contents

# Contributors

*M. Ray Denny* • Michigan State University, East Lansing, MI 48824-1117

*Gerald I. Dewey* • University of Illinois, Champaign, IL 61820

*Peter W.D. Dodd* • Bell Northern Laboratories, Ottawa, Ontario, Canada K17 4H7

*Douglas S. Grant* • University of Alberta, Edmonton, Alberta, Canada T6G 2E9

*Werner K. Honig* • Dalhousie University, Halifax, Nova Scotia, Canada B3H 4J1

*Donald F. Kendrick* • Middle Tennessee State University, Murfreesboro, TN 37132

*Douglas L. Medin* • University of Illinois, Champaign, IL 61820

*Julie J. Neiworth* • Michigan State University, East Lansing, MI 48824-1117

*Mark E. Rilling* • Michigan State University, East Lansing, MI 48824-1117

*H.L. Roitblat* • University of Hawaii at Manoa, 2430 Campus Road, Honolulu, HI 96822

*Stephen F. Sands* • University of Texas at El Paso, El Paso, TX 79902

*Peter J. Urcuioli* • Purdue University, Lafayette, IN 47907

*Edward A. Wasserman* • The University of Iowa, Iowa City, IO 52242

*R.G. Weisman* • Queens University, Kingston, Ontario, Canada K7L 3N6

*Anthony A. Wright* • University of Texas Health Science Center, Graduate School of Biomedical Sciences, 6420 Lamar-Fleming Blvd., Houston, TX 77030

# Preface

This volume was conceived during the summer of 1980. Mark and I had returned from The Third Harvard Symposium on Quantitative Analysis of Behavior: Acquisition, and from Spear and Miller's Binghamton, NY, Conference on Information Processing in Animals: Memory Mechanisms. Our summer discussions of what we learned at Harvard and SUNY often involved Ray Denny, who for several years had been concerned with the transition of animal learning from a strict stimulus-response (S-R) approach to a more cognitive approach. This volume was conceived from these discussions.

We noted that nearly everyone's research was guided by some perspective or theoretical framework based partly on a combination of research results and individual opinions about what animals can do. This volume was thus conceived as a collection of chapters in which animal memory researchers could publicly state their opinions about animal memory, with little concern for substantiating them with test data. This volume would provide an opportunity for researchers to reveal their private thoughts on animal memory. However, when one's private notions are examined carefully, as happens in the transfer from thought to paper, they are altered, constrained, viewed more clearly, and ultimately become theoretical formulations. Thus, what began as an attempt to publish the private perspectives that guide the research of a dozen investigators is now the theoretical formulations of nine animal-memory researchers. We do hope that these chapters will have a guiding influence on the re-emerging field of comparative cognition, and that our readers will be stimulated to examine their own perspectives and guiding influences.

This volume is organized in three main sections of three chapters each. The first section, The Grand Approach, is a collection of chapters with a meta-theoretical perspective. Roitblat and Weisman, in Chapter 1, describe the Tactics of Comparative Cognition. They see comparative cognition as a sub-field of cognitive science bridging the gap between neuroscience and human cognition. They also see it as a biological approach to the study of knowledge. A major part of their chapter covers the antagonistic relationship between animal cognition and behaviorism. Ray Denny, in Chapter 3, expresses a behaviorist's view of animal cognition and demonstrates how the "cognitive invasion" can be assimilated into stimulus-response psychology thus benefiting both approaches to the study of animal and human behavior. Rilling and Neiworth, in Chapter 2, view animal cognition in the context of human cognition and propose that animal and human cognition studies demonstrate the viability of a general processes approach to cognitive behavior.

The second section, Memory Processes, presents three chapters concerned with the processes, properties, and mechanisms of short-term memory in animals. Ed Wasserman, in Chapter 4, is concerned with the dualism of prospection and retrospection whereas Honig and Dodd, Chapter 5, are concerned with the specific mechanisms of prospection. Wasserman presents Konorski's original analysis of short-term memory, evaluates it in view of current evidence and suggests further research of the conditions and variables that determine the relative roles of prospection and retrospection in short-term memory. Honig and Dodd suggest that animals are capable of anticipating future stimulus events and of forming intentions of how to respond to them. They review in detail evidence favoring prospection and build a theoretical framework of the mechanisms and processes of prospective working memory. In Chapter 6, Wright, Urcuioli, and Sands argue that one of the main obstacles to obtaining data on animal memory is proactive interference. Using a small number of stimuli in memory tasks is the primary reason animals' memories appear so inferior to humans' memories. They further demonstrate that increasing the number of stimuli used reduces the animal-human differences.

The third section, Theoretical Issues, presents two highly developed theories of animal memory, one based on pigeon short-term memory experiments (Kendrick & Rilling, Chapter 7), and one based on delayed alternation in the rat (Grant, Chapter 8). Kendrick and Rilling propose a theory mostly derived from viewing animal short-term memory from the perspective of human memory theories. They argue that the animal data support current views of process-based active and inactive representations rather than structural-based two-stores theories of short- and long-term memory. Grant also presents a detailed account of the animal memory system in which memories are multi-dimensional, active or inactive, retrieved and rehearsed, tagged, updated, and discriminated. His model of memory is then evaluated in terms of the results of studies of delayed alternation in the rat. Medin and Dewey contribute the final chapter whereby they consider the past, the present, and the future of theories of animal memory. Their chapter nicely sums up where we have been, where we are, and where we should be headed.

There are many people to thank for their role in this book. In addition to the contributors, we would like to especially note the invaluable assistance of Jack Burton and Kathryn Kolwicz of Lawrence Erlbaum Associates. If we had been better editors, their jobs would have been easier. We also acknowledge the secretarial assistance of Carolyn Powers and Nancy Dickson. This book was made possible, in part, by a Grant from the National Science Foundation to Mark Rilling (BNS 81-10966), and by a Postdoctoral Fellowship from the National Institute of Mental Health awarded to Donald (Skip) Kendrick (1 F32 MH08928-01).

*Donald (Skip) Kendrick*
*Mark Rilling*
*Ray Denny*

# I THE GRAND APPROACH

# 1 Tactics of Comparative Cognition

H. L. Roitblat
*University of Hawaii at Manoa*

R. G. Weisman
*Queen's University at Kingston*

Perhaps the most striking feature of living organisms is their responsiveness to environmental events. Philosophers and scientists have sought the mechanisms underlying this responsiveness throughout written history, generating many scientific disciplines, including physiology, biochemistry, neuroscience, and psychology. This book presents a psychological inquiry into the role of cognitive processes in determining the responsiveness of animals to their environment. Comparative cognition brings ethological, evolutionary, neural, and psychological perspectives to the understanding of behavior. Although this chapter develops a distinctly psychological approach to comparative cognition, proper development of the field requires contributions from each of the other disciplines.

Comparative cognition is a subfield of cognitive science, serving the more general discipline as a bridge between neuroscience and human cognition and as an independent and distinctly biological approach to the study of the structure and use of knowledge. Until recently, research in neuroscience has been hampered by an impoverished understanding of animal cognition that, in turn, has restricted the types of neural mechanisms investigated and raised unnecessary doubts about the relevance of a neuroscience of nonhumans to the neuroscience of humans (Norman, 1973).

This chapter views animals as what might be called, following Simon (1981), "thinking organisms": intelligent animals capable of adapting to their environments through the expression of varied cognitive skills. These skills include, for example, learning, remembering, problem solving, rule and concept formation, perception, and recognition. Analysis in comparative cognition seeks to explain behavior in terms of these skills (and the representations and processes that instantiate them) as they interact with the environment. However complex, the cognitive systems of the thinking organism are physical entities, determined by the phylogenetic and ontogenetic history of the organism. Comparative cognition can also be viewed as the part of comparative biology concerned with the evolution of the mind. It seeks to understand the evolutionary and functional bases of cognitive mechanisms.

Finally, comparative cognition is more than a subject matter and a method, it is a perspective. Comparative cognition organizes thinking about animal behavior. It is, in Kuhn's (1970) sense, a paradigm or disciplinary matrix, and, in Lakatos's (1970) sense, a research program. The function of a research program is to provide a framework for theory and a heuristic for research.

Research programs can not be falsified, but they must be criticized. Scientific criticism also actively confronts competing theories, derived from a research program, with relevant data. The tactics of strong inference (Platt, 1964) provide the means to conduct this confrontation between a wealth of conceivable models of the cognitive competencies and the behaviors they purport to explain.

## COMPARATIVE COGNITION AND BEHAVIORISM

The last decade has seen a remarkable change in the way many psychologists think about animal behavior. Between 1930 and 1970, behaviorism was the dominant paradigm in comparative psychology. Many of the founding assumptions of comparative cognition are a reaction to the perceived inadequacy of behaviorist assumptions about the responsiveness of animals. A comparison of the contrasting assumptions of behavior theory and comparative cognition provides a historical context for the current status of comparative cognition.

According to the behaviorist point of view (behavior theory), the appropriate task for psychology is solely the study of behavior (e.g. Skinner, 1950, 1974, 1977; Wessells, 1982). The purpose is to discover lawful relations between experience and action. Environmental events are said to play the key role in determining the behavior of organisms; the source of the action is in the environment, not the individual. The only appropriate level of description is in terms of strict observables, i.e., physically specifiable stimuli and responses. Given a sufficiently detailed description of the environment and of the lawful relationships between environmental and behavioral variables, one can understand all behavior. Understanding is indexed solely by the ability to predict and control behavior.

According to behavior theory, the development of a full account of behavior does not require an understanding of mental events. Although internal events may exist, their analysis is irrelevant. If they exist, they consist of private stimuli and sometimes private responses, which may "occur on a scale so

small that [they] cannot be detected by others [Skinner, 1974, p. 103]."
Furthermore, their cause is in the environment, so we can go directly to these
"prior physical causes while bypassing feelings or states of mind . . . If all
linkages are lawful, nothing is lost by neglecting the supposed nonphysical
link [Skinner, 1974, p. 13]." Or, to restate the behaviorist viewpoint, mental
events are epiphenomenal to the determination of behavior because the situation acts directly to determine behavior. "From a behaviorist point of view
'causes' produce their effects across a temporal gap, and in many cases that gap
can be large. Thus to be comfortable with a behavioristic position one must be
comfortable with the notion of 'action at a distance' [Branch, 1982, p. 372]."
In short, the behaviorist finds the internal states of the organism irrelevant in
descriptions of behavior; experience works its effects directly on later behavior
with nothing, at least nothing important, occurring between.

Comparative cognition rejects these and other fundamental assumptions of
behavior theory including: (1) the nonphysicality of mental events; (2) the
irrelevance of mental events; (3) the adequacy of substituting functional
analysis for theory; and (4) the simplicity and centrality of instrumental
learning and classical conditioning. First, Skinner's (1974) association of
mental events with nonphysical events reveals his misunderstanding of modern
cognitive science. It is not clear to us what a nonphysical event could be, but
comparative cognition is firmly committed to the physical, specifically biological, nature of mental events. We view cognitive processes and structures as
physical entities analogous to the biological processes and structures forming
the explanatory basis of modern genetics. A psychology that ignores cognition
is, we contend, as irrelevant as a biology that ignores the gene.

Second, the behaviorist claim that mental events are irrelevant is equally
unsatisfactory. A complete analysis based on functional analysis of relationships between stimuli and responses is possible only if mental events are
linked to environmental and behavioral events in one way. The behaviorist
must assume strict isomorphism between stimulus events and mental events
(Bolles, 1975): Given a specified stimulus event, one and only one mental
event must result, and given that mental event, one and only one behavioral
event must follow. By this analysis, behaviorism has a theory of cognitive
structure, albeit a simple one, consisting of connections between internal
representations of stimuli and responses. While parsimony is often a virtue in
science, oversimplification is not. The behaviorist model of internal representation provides a false sense of parsimony, because its hidden assumptions do
violence to the complexity of the animal mind, leaving the behaviorists with
an impoverished, uncritical view of mental structure.

Third, by emphasizing functional relationships between stimuli and responses, behaviorism ignores activities traditionally associated with successful explanation and understanding in science: model and theory building. Of
course, the behaviorist has theories (Skinner, 1974), but these most often lie

unarticulated, buried in descriptions of lawful relationships, and are, therefore, resistant to critical appraisal.

Fourth, the behaviorist position has led animal psychologists to an obsession with "simple" learning processes, in the vain hope that an understanding of these would lead automatically to an understanding of "complex" processes. Even the simple processes have been resistant to purely behavioristic explanation and have become rich domains for cognitive analysis (e.g., Dickinson, 1980; Wagner, 1981). Furthermore, one does not need to deny the importance of classical conditioning and instrumental learning to insist it is a serious error to ignore the rich variety present in the behavior of organisms. Animals perceive complex spatial relationships, represent the order of events, time the durations of events, predict the results of ballistic motion, and much more, but behaviorism must remain silent on how animals do these wondrous things. Comparative cognition seeks to explore this complexity directly, because no simple additive combination of independently discoverable simple processes is likely to explain it. If one wishes to understand the complexity of the animal mind, then one must confront that complexity.

The essential difference between the cognitivist and behaviorist positions is the cognitivist emphasis on the organism as opposed to the behaviorist emphasis on the relationship between situations and behaviors. In place of the behaviorist's "action at a distance" (Branch, 1982), the cognitivist asserts that if an organism's past experience is to change its future behavior, then some change (or succession of changes) must occur within the organism. The available evidence from comparative cognition indicates that the representations used by animals are more complex than the simple S-R (stimulus-response) associations of behavior theory, whether considered as Hullian habits or as Skinnerian functional relations. Obviously, mental events are lawfully related to experience, but they are not limited by strict isomorphism with either stimuli or responses (see Bolles, 1975; Dickinson, 1980; Roitblat, 1982). If mental representations are not isomorphic with external events, then their description becomes at once interesting and complex. Instead of only one possible representational system for all tasks, there are many possible systems for each task. The research problem for comparative cognition is to determine which of these mental systems best explains an animal's behavior in the task.

In the early stages of the development of comparative cognition, some thought it expedient to divide behaviors into those that are "cognitive" and those that are not. Some behaviors were to be explained by behavior theory; other more complex behaviors, not easily explainable in those terms, were to be explained by cognitive principles. These complex behaviors were taken as evidence that animals did think, after all (Terrace, 1981, 1983). We believe this attempt to partition behavior is mistaken. Cognition is not the portion left over when behaviorist explanations have been exhausted (see Bever, 1983). Attributing behaviors to allegedly simple S-R type explanations does not make them noncognitive. S-R relations are only one kind of representation an organism might use. The application of labels—e.g., classical conditioning and generalization—to behavioral phenomena gives the impression that they have been explained in simple terms. The concepts so labeled are, as we have noted already, not primitives, but themselves complex phenomena begging for explanation.

While the fall of behaviorism and the rise of cognitive science is a scientific revolution (Kuhn, 1970), much of the experimental rigor associated with the behaviorist tradition has become part of comparative cognition as well. Careful demonstrations of the lawfulness and predictability of behavior must necessarily accompany any serious attempt to explain it. Such demonstrations are often provided by the same careful investigators who formerly conducted behavioral research.

## FIRST PRINCIPLES IN COMPARATIVE COGNITION

### Comparative Cognition as Cognitive Science

Comparative cognition is concerned with how animals gain and use knowledge about their world. Behavior is seen, not as the primary target for investigation, but as an indicator of an organism's knowledge, as an ambassador (Romanes, 1883/1898; Wasserman, 1983). Although in cognitive science the study of behavior is a means to an end, it is not a trivial means. Sound inferences about cognition can not be gained from infirm behavioral ambassadors. The cognitivist requires good behavioral data as much as the behaviorist.

The approach assumes organisms have internal cognitive structure (i.e., minds) consisting of psychological processes analyzed conceptually into a series of stages or states. Each stage takes time and operates on the output of the previous stage, transforming the output, and perhaps combining it with information from other sources. An example of the concept of stages of processing is the notion of retrospective and prospective processing in the delayed discrimination task (Honig & Thompson, 1982; Honig & Wasserman, 1981; Roitblat, 1980). In this instance, a prospective decision process is inferred to operate on the perceptual products of an earlier retrospective memory process. For example, in at least some instances of delayed conditional discrimination, pigeons can operate on their representation of the sample stimulus to produce a prospective code specifying the correct response and its consequences (Grant, 1982; Peterson, 1983; Roitblat, 1980; Roitblat, 1982).

We think that prospective representation is a useful concept, however, nothing in the concept of stages confines one to a linear model of processing. An interactive model would include both sensory-driven (bottom-up) and concept-driven (top-down) stages of processing. Sometimes a purely linear model may be in order; in other circumstances, stages may operate in parallel or through feedback of partial information influencing how subsequent states process later information. Also, the stage concept need not imply the stages are always discrete, easily separable from one another. It is probably more appropriate to view stages as "snapshots" of the state of an animal's cognitive system at successive points in time during a task. Finally, the contents and order of cognitive stages should not be viewed as static. Experience may subject the mind of the thinking organism to dynamic reorganization.

*Representation.*    External events do not enter the mind directly. One can not conceptually "push" stimuli into animals' minds. Thus animals can not associate external events directly, one with another. Rather, external events must be encoded symbolically; they must be represented. Representation is a fundamental concept in cognitive science (see Roitblat, 1982). The study of representations and the processes that generate them is the chief experimental work of cognitive science. We seek to study the representation and processing of knowledge at each stage in cognition, the causal order of the stages, and how environmental factors contribute to the determination of those stages, processes, and representations.

Thinking and consciousness have long been problematic terms for psychology, and for comparative cognition in particular. There is no general agreement about what these terms mean, but however defined they refer to something occurring within the head of the thinking organism. The task for cognitive scientists is to describe and explain that something, whatever one might want to call it. It is not the existence of minds (or thought and consciousness) that explains behavior, it is the *particular* active contents of those minds that do. One must ask why a particular representation is active and not others, and how has it come to function as it does?

Purposive action, goal directed and motivated behavior, is the unit of analysis in comparative cognition. Cognition is goal oriented. Not all behavior is problem solving but comparative cognition is especially interested in this aspect of animal behavior. Cognitive processes and representations are not independent of the problems that they solve. For example, the representation of order information can not be understood independently of the task in which it is acquired and used. Thus, order tasks that require sequence production (e.g., Straub, Seidenburg, Bever, & Terrace, 1979) may be represented quite differently than tasks that require only sequence recognition (e.g., Weisman, Wasserman, Dodd, & Larew, 1980). The problem and the intelligence of the problem solver jointly determine cognition (Simon, 1981). Put simply, the cognitivist seeks to understand how animals use their intelligence to solve problems in their environment.

## Comparative Cognition as Comparative Biology

A mistaken view of cognitive evolution is that it results in the addition of new layers of cognitive skills to those already laid down, similar to the layers of an onion. Someone with this view might describe a chimpanzee as a human being lacking precisely those mental structures necessary for the development of language. In other words, take the language mechanism from a human and the result would be a chimp (after making minor adjustments for the indirect kinship between the two species). It can be argued that recent attempts to teach language to chimpanzees, whatever their success or failure, exploit the onion analogy. The interaction between old and new structures in the evolving brain is considerably more dynamic than this analogy suggests. A change to one part of the brain through evolution may result in a virtual reorganization of many other parts of the brain as well. Even within an animal's lifetime, trauma or surgical intervention to one brain structure can result in functional reorganization of the brain (e.g., Goldman, 1978). This argument has two implications. First, any attempt at comparative psychology based on a linear ordering of species—or on the accumulation of cognitive capacities of ever increasing power—is doomed to failure. Second, comparative cognition is not a conceptually easy undertaking. Phylogenetic change may have far reaching consequences, making the development of comparative principles a difficult, but hardly impossible, undertaking.

Darwin (1871/1920) and Romanes (1883/1898) asserted mental evolution, no less than physiological and anatomical evolution, was a continuous process. By this they meant that no species, including humans, developed in total isolation from the others. Each species developed from ancestral species and exhibited continuity with its ancestors by bearing many of their characteristics. This is not the place for a discussion of speciation (see Dawkins, 1976; Wilson, 1976), but this argument does not imply that any differences among species are simply matters of degree. Consider an example from anatomy. First, although birds' wings are related to parts of the horses' foot and to parts of the human hand, they are not identical organs. Second, although bat wings and bird wings resemble one another, they are not identical organs either. The first relationship is one of homology. These structures resemble one another because sometime, long ago, both the horse and the bird had a common ancestor (presumably the ancestor of all vertebrates). Differences between the bird's wing and the horse's foot are the product of divergent evolution. The second relationship is an example of analogy. Whatever similarities exist between bat and bird wing result from the similar functions both perform. An organ of flight must have certain aerodynamic properties, whatever the ancestry of its possessor. Similarities between the bird's wing and the bat's wing are the product of convergent evolution.

The concepts of homology and analogy also have application in comparative cognition. Species may share cognitive competencies because they have a common ancestor, or they may have similar competencies because both species have adapted to the environment in the same way. When similar cognitive skills develop through convergent evolution, the environment can be seen to play a large role in determining those competencies. This argument is related to Gibson's (1966) theory regarding the role of the environment in perception. Certain of the properties of classical conditioning, for example, are seen universally in all the species studies. One could argue that these universal properties are the product of a universal conditioning mechanism, evolved once and inherited by most known organisms (with the occasional embellishment). On the other hand, the possibility exists that the universal properties of conditioning are not an homology, but a response to relationships that underlie causation in the environment. According to this view, universality exists because all animals face the same kinds of causal relations and have, via convergent evolution, developed the necessary cognitive machinery to deal effectively with causation. Without committing oneself to divergent or convergent evolution of classical conditioning, one can ask such questions about conditioning and cognitive competencies more generally. Attempts to distinguish between homology and analogy in the evolution of cognition are often difficult, requiring careful observation of well-chosen groups of species.

## COMPARATIVE COGNITION AND THE PHILOSOPHY OF SCIENCE

### The Logic of Science

Any given observation, in any area of science, is necessarily consistent with many alternative explanations. In comparative cognition the best one can do is to manipulate the features of situations (inputs) and observe the resulting behaviors (outputs). From these input/output relationships one infers the functions of the internal "information processing machine." If one attempts to open the machine (i.e., the skull of the animal) to get a better view, one finds only smaller machines, and so forth. Moore (1956) has shown that given any particular machine and any experiment of arbitrary complexity, including a series of experiments of finite length, there exist other machines that could give exactly the same results under the same circumstances. Whatever properties one discovers for this machine, those same properties would be included in another machine as well. While one may be able to discriminate any two particular machines by their input/output properties (hence the value of strong inference techniques), no experiment or series of experiments is sufficient to identify (i.e., prove) a given machine (i.e., theory) from among the set of all possible machines.

All sciences are indeterminate in exactly the manner Moore (1956) described. This form of indeterminacy is not a special problem for cognitive science, but a problem for all scientific enterprises. On logical grounds, no scientific theory is certain and none can be justified. The "machines" of physics are as indeterminant as those of psychology. Positivists and other verificationists regard a theory as justified if the available evidence is consistent with the theory. Theories go beyond their observations in three ways, however. First, a theory is a universal extension of a finite number of observations. Second, a theory is a precise formulation based on observations accurate only within the limits of experimental error. Third, a theory attempts to specify the mechanisms underlying observed phenomena, but the phenomena themselves are not those mechanisms. In short, theories transcend their observational justification in universal extension, precision, and depth. None of these extensions can be justified (i.e., proved true). One possible solution, that taken by positivists (Watkins, 1978), is to sacrifice depth by removing the assertion that certain mechanisms underlie the observations, constructing theories only at the level of the observable. This is the route taken by Skinner (1950), for example. His attempt to de-ontologize theory has met with only limited success even on its own terms: Consider the interminable conflict over matching versus maximizing as fundamental (Baum, 1981; Herrnstein, 1970; Hinson & Staddon, 1981).

Positivism fails on more general considerations, because positivist theories still transcend the observations in two other ways. Finite observations can not be justifiably extended (i.e., with certainty) to universal statements. No matter how many observations are used as the basis of the universal statement, a later observation could produce inconsistent data. One attempt to remedy this restriction involves a substitution of probable for certain (Lakatos, 1970). Consistent observation, it was asserted, increased the probability that a given theory was true. The main objection to this view is that the probability of a hypothesis, given a set of observations, is inversely proportional to its content and its specificity; therefore, any universal hypothesis must have a probability of zero (Popper, 1962).

The nonjustificationist approach to scientific method attempts to remedy these difficulties by sacrificing certainty instead of depth. Although this position underlies much of cognitive science (Lachman, Lachman, & Butterfield, 1979), in the present context we shall do no more than sketch its essential features. They are: (1) In contrast to positivism, nonjustificationism does not claim that it is possible to apportion a scientific theory into theory-free and theory-laden statements. Observations are recognized to be dependent on the theory held by the observer. Not even the data on which theories are based are infallible. (2) Scientific theories can not be proved. Certainty of either observation or theory is unattainable. (3) Scientific theories can not be disproved. All theories specify the conditions appropriate for valid observation. If those specifications are not met, then failure to make the appropriate observation can not count as evidence against the theory. As a trivial example, consider the positive statement that all swans are white.. The discovery of a black swan-like bird does not disprove the statement (that all swans are white), because the positive statement that the black bird in question is a swan can not be proved. Whatever tests one performs consistent with the black bird being a swan, the next test may be contrary (cf. Moore's [1956] theorem). Nonjustificationism asserts that the proper unit of scientific endeavor is the research program (Lakatos, 1970), which consists of a set of eternally conjectural propositions and the relevant data. Scientific progress is measured by the degree to which a research program's theories lead to the discovery of novel facts (a positive heuristic) and provide increased (progressive) explanatory capacity. Science consists of the critical comparison of research programs in light of observation and fact.

It follows from this position that comparative cognition is not subject to ultimate justification or falsification, but eternally subject to criticism. Data can not prove the cognitive position, but they can provide evidence of its positive heuristic and progressive nature. At present, this particular program appears to provide the strongest heuristic. It will be replaced, nevertheless, when the flow of new facts and understanding slows and another research program with a stronger heuristic is developed.

## The Tactics of Science

We have argued that comparative cognition is subject to the same general constraints and limitations as every other area of science. In addition to these general limitations, scientists working in comparative cognition must deal with an immense range of species and tasks as they seek explanations among nearly infinite supply of model structures and processes. It is clear that the study of comparative cognition will not be easy. Still, it is better to face the complexity of the discipline squarely than to be bound by simplistic thinking. The best one can do is to formulate bold conjectures and test them, ruthlessly, one against another in light of the data. Science, and comparative cognition in particular, is no place for intellectual hegemony. Long ago, Chamberlin (1904) counseled against allowing a single theory to rule ones research. Yet, a precious small number of theories have ruled research in animal learning for decades. Good science relies on the art of proposing exciting alternative explanations to be tried by clever observations and experiments. Useful advice on the relentless use of criticism in the service of scientific progress is provided by Chamberlin (1904), Lakatos (1970), Platt (1964), and Popper (1962).

This is tough talk, stronger than our actions. Scientists, it seems, fare little better than other humans at using strong inference, i.e., the logic of alternative working hypotheses and their test. Scientists often fail to: (a) seek disconfirmitory evidence (even when it is available), (b) test alternative hypotheses, or (c) consider whether evidence supporting a favored hypothesis will support other hypotheses as well (Tweney, Doherty, & Mynatt, 1981). We scientists bear the weaknesses of our species. Our human frailties, however, do not invalidate the logic of the method, but only bring to mind how difficult it is to do good science. If we are up to the challenge, it is better to follow sound advice imperfectly than to eschew it as too exacting.

Behaviorism led ultimately to a fixation with conditioning and simple forms of instrumental learning. Students of comparative cognition have investigated a wide variety of tasks and species. A sample list (see Roitblat, Bever, & Terrace, 1984) of the tasks includes concept learning, delayed matching to sample, memory for food caches and spatial locations, sequence discrimination and production, and timing, studied in species as diverse as bluejays, chimps, Clark's nutcrackers, dolphins, gerbils, hummingbirds, marshtits, monkeys, starlings, parrots, as well as the "traditional" laboratory species of pigeons and rats. The success of comparative cognition depends on the expansion of this diversity. There really is, we believe, an advantage to a pluralistic approach to cognition both in terms of tasks and species. Without this diversity, it is impossible to assess the extent to which our theories are ad hoc and our models based on hidden assumptions.

## Dependent Variables

Skinner (1950) advised behaviorists to adopt rate of responding as their dependent variable and to avoid the use of other measures of behavior. In comparative cognition, we appreciate the virtues of diversity in dependent variables. We shall describe some useful dependent variables as examples, not prescriptions.

*Frequency Correct.*    In choice procedures, the proportion of responses correct or incorrect is tallied. This measure is often used in studies of animal memory (e.g., Roberts & Grant, 1976). As strength or clarity of memory decreases so should the frequency of the correct response.

*Analysis of Response Protocols.*    After classification as correct or incorrect, many responses can be further classified to permit more detailed analysis (e.g., Bishop, Fienberg, & Holland, 1975). For example, the same system that produces correct responses also produces errors. The pattern of errors and the relationship among different sorts of errors can be a source of useful information. Roitblat and Scopatz (1983) provide an example of the application of this method in animals.

*Rate of Response.*    Not all experiments in comparative cognition involve choice, some use the method of successive (go/no go) discriminations, for example. In these latter experiments, the main dependent variable is the rate of responding. Nelson and Wasserman (1978) describe some useful successive delayed discrimination procedures. Rate of response in these experiments is logically related to frequency correct in choice experiments, in the same way that theories of signal detection (Egan, Schulman, & Greenberg, 1959) and cognitive integration (Anderson, 1981) relate choice and rating measures of decision making.

Choice responses reflect the application of a binary decision criterion to some underlying continuous rating of preference, while response rate reflects the rating of preference more directly. Response rate also reflects animals' strong biases toward responding in go/no go discriminations. Experiments using frequency of correct response and rate of response measures for choice and successive discriminations complement one another, strengthening the inferences they jointly support.

*Chronometric Analyses.*    The time it takes to perform a task can be used as a dependent measure. Time is used either as an index of the processing complexity required by a task (Sternberg, 1969) or as a measure of the strength of the outcome instruction or expectancy (Hulse, 1978). As an index of processing complexity, increased response latencies indicate more steps taken by the system. Through various manipulations of the features of a task (and of the stimuli involved in a task), it is sometimes possible to infer the functions of various stages in the process (Sternberg, 1975).

## SUMMARY

Comparative cognition is a subfield of cognitive science and at the same time part of comparative biology. Its goal is to understand living, evolving cognitive systems. It is comparative in at least three senses: (1) in the comparison between animal with humans minds; (2) in its concern for the evolution of intelligence and mind; and (3) as a bridge between neuroscience and cognitive science. Behaviorism rejected, as useless, the proposition that animals have minds. Comparative cognition accepts that proposition as given, and asks what kinds of minds do animals have? Representation is a fundamental concept in cognitive science. The task of comparative cognition is to infer representations and processes and their stages of occurrence in the mental competencies of animals, and to test alternative models of these competencies one against another. In the quest to understand the minds of animals, comparative cognition seeks not certainty, but depth of explanation.

## ACKNOWLEDGMENTS

The preparation of this chapter was supported by Grant BNS 82–03017 from the US National Science Foundation and Grant 1RO1–MH37070 from the US National Institute of Mental Health to Herbert Roitblat and by Grant AO182 from the Natural Science and Engineering Council of Canada to Ronald Weisman.

## REFERENCES

Anderson, N. H. *Foundations of information integration theory.* New York: Academic Press, 1981.

Baum, W. M. Optimization and the matching law as accounts of instrumental behavior. *Journal of the Experimental Analysis of Behavior,* 1981, *36,* 387–403.

Bever, T. G. On the road from behaviorism to rationalism. In H. L. Roitblat, T. G. Bever, & H. S. Terrace (Eds.), *Animal cognition.* Hillsdale, NJ: Lawrence Erlbaum Associates, 1984.

Bishop, T. M. M., Fienberg, S. E., & Holland, R. W. *Discrete multivariate analysis: Theory and practice.* Cambridge, MA: MIT Press, 1975.

Bolles, R. C. Learning, motivation, and cognition. In W. K. Estes (Ed.), *Handbook of learning and cognitive processes.* Hillsdale, NJ: Lawrence Erlbaum Associates, 1975.

Branch, M. N. Misrepresenting behaviorism. *The Behavioral and Brain Sciences,* 1982, *5,* 372–373.

Chamberlin, T. C. The methods of the earth sciences. *Popular Science Monthly,* 1904, *66,* 66–75.

Darwin, C. *The descent of man and selection in relation to sex.* New York: Appleton, 1920. (Originally published, 1871.)

Dawkins, R. *The selfish gene.* Oxford: Oxford University Press, 1976.

Dickenson, A. *Contemporary animal learning theory.* New York: Cambridge University Press, 1980.

Egan, J. P., Schulman, A. I., & Greenburg, G. Z. Operating characteristics determined by binary decisions and by ratings. *Journal of the Acoustical Society of America,* 1959, *31,* 768–773.

Gibson, J. J. *The senses considered as perceptual systems.* Boston: Houghton-Mifflin, 1966.

Goldman, P. S. Neuronal plasticity in primate telecephalon: Anomolous projections induced by prenatal removal of frontal cortex. *Science,* 1978, *202,* 768–770.

Grant, D. S. Prospective versus retrospective coding of samples of stimuli, responses, and reinforcers in delayed matching with pigeons. *Learning and Motivation,* 1982, *13,* 265–280.

Herrnstein, R. J. On the law of effect. *Journal of the Experimental Analysis of Behavior,* 1970, *13,* 243–266.

Hinson, J. M., & Staddon, J. E. Maximizing on interval schedules. In C. N. Bradshaw (Ed.), *Recent developments in the quantification of steady-state operant behavior.* New York: Elsevier North-Holland, 1981.

Honig, W. K., & Thompson, R. A. Retrospective and prospective processing in animal working memory. In G. H. Bower (Ed.), *The psychology of learning and motivation.* New York: Academic Press, 1982, *16,* 239–283.

Honig, W. K., & Wasserman, E. A. Performance of pigeons on delayed simple and conditional discriminations under equivalent procedures. *Learning and Motivation,* 1981, *12,* 149–170.

Hulse, S. H. Cognitive structure and serial pattern learning by animals. In S. H. Hulse, H. Fowler, & W. K. Honig (Eds.), *Cognitive processes in animal behavior.* Hillsdale, NJ: Lawrence Erlbaum Associates, 1978.

Kuhn, T. S. *The structure of scientic revolutions* (2nd ed.). Chicago, IL: University of Chicago Press, 1970.

Lachman, R., Lachman, J. L., & Butterfield, E. R. *Cognitive psychology and information processing.* Hillsdale, NJ: Lawrence Erlbaum Associates, 1979.

Lakatos, I. Falsification and the methodology of scientific research programs. In I. Lakatos & A. Musgrave (Eds.), *Criticism and the growth of knowledge.* London: Cambridge University Press, 1970.

Moore, E. F. Gedanken-experiments on sequential machines. In C. E. Shannon & J. McCarthy (Eds.), *Automata studies.* Princeton: Princeton University Press, 1956.

Nelson, K. R., & Wasserman, E. A. Temporal factors influencing the pigeon's successive matching-to-sample performance: Sample duration, intertrial interval, and retention interval. *Journal of the Experimental Analysis of Behavior,* 1978, *8,* 153–162.

Norman, D. A. What have animal experiments taught us about human memory? In J. A. Deutsch (Ed.), *The physiological basis of memory.* New York: Academic Press, 1973.

Peterson, G. B. The differential outcome procedure: A paradigm for studying how expectancies guide behavior. In H. L. Roitblat, T. G. Bever, & H. S. Terrace (Eds.), *Animal cognition.* Hillsdale, NJ: Lawrence Erlbaum Associates, 1984.

Platt, J. R. Strong inference. *Science,* 1964, *146,* 347–353.

Popper, K. *Objective knowledge.* Oxford: Oxford University Press, 1962.

Roberts, W. A., & Grant, D. S. Studies of short-term memory in the pigeon using the delayed matching to sample procedure. In D. L. Medin, W. A. Roberts, & R. T. Davis (Eds.), *Processes of animal memory.* Hillsdale, NJ: Lawrence Erlbaum Associates, 1976.

Roitblat, H. L. Codes and coding processes in pigeon short-term memory. *Animal Learning and Behavior,* 1980, *8,* 341–351.

Roitblat, H. L. The meaning of representation in animal memory. *The Behavioral and Brain Sciences,* 1982, *5,* 353–406.

Roitblat, H. L., Bever, T. G., & Terrace, H. S. (Eds.). *Animal cognition.* Hillsdale, NJ: Lawrence Erlbaum Associates, 1984.

Roitblat, H. L., & Scopatz, R. A. Sequential effects in pigeon delayed matching-to-sample performance. *Journal of Experimental Psychology: Animal Behavior Processes,* 1983, *9,* 202–221.

Romanes, C. J. *Mental evolution in animals.* New York: Appleton, 1898. (Originally published, 1883.)

Simon, H. A. *The sciences of the artificial* (2nd ed.). Cambridge, MA: The MIT Press, 1981.

Skinner, B. F. Are theories of learning necessary? *Psychological Review,* 1950, *57,* 193–216.

Skinner, B. F. *About behaviorism.* New York: Knopf, 1974.

Skinner, B. F. Why I am not a cognitive psychologist. *Behaviorism,* 1977, *5,* 1–10.

Sternberg, S. Memory-scanning: Mental processes revealed by reaction-time experiments. *American Scientist,* 1969, *57,* 421–457.

Sternberg, S. Memory Scanning: New findings and current controversies. *Quarterly Journal of Experimental Psychology,* 1975, *27,* 1–32.

Straub, R. O., Seidenberg, M. S., Bever, T. G., & Terrace, H. S. Serial learning in pigeon. *Journal of the Experimental Analysis of Behavior,* 1979, *32,* 137–148.

Terrace, H. S. Animal versus human minds. *The Behavioral and Brain Sciences,* 1982, *5,* 391–392.

Terrace, H. S. Animal cognition. In H. L. Roitblat, T. G. Bever, & H. S. Terrace (Eds.), *Animal cognition.* Hillsdale, NJ: Lawrence Erlbaum Associates, 1984.

Tweney, R. D., Doherty, M. E., & Mynatt, C. R. *On scientific thinking.* New York: Columbia University Press, 1981.

Wagner, A. R. SOP: A model of automatic memory processing in animal behavior. In N. E. Spear & R. R. Miller (Eds.), *Information processing in animals: memory mechanisms.* Hillsdale, NJ: Lawrence Erlbaum Associates, 1981.

Wasserman, E. A. Animal intelligence: Understanding the minds of animals through their behavioral "ambassadors." In H. L. Roitblat, T. G. Bever, & H. S. Terrace (Eds.), *Animal cognition.* Hillsdale, NJ: Lawrence Erlbaum Associates, 1984.

Watkins, J. The Popperian approach to scientific knowledge. In G. Radnitzky & G. Anderson (Ed.), *Progress and rationality in science.* Dordrecht, Holland: Reidel, 1978.

Weisman, R. G., Wasserman, E. A., Dodd, P. W. D., & Larew, M. B. Representation and retention of two-event sequences in pigeons. *Journal of Experimental Psychology: Animal Behavior Processes,* 1980, *6,* 312–325.

Wessells, M. G. A critique of Skinner's views on the obstructive character of cognitive theories. *Behaviorism,* 1982, *10,* 65–84.

Wilson, E. O. *Sociobiology.* Cambridge, MA: Harvard University Press, 1976.

# 2 Comparative Cognition: A General Processes Approach

Mark E. Rilling
Julie J. Neiworth
*Michigan State University*

The cognitive approach is the dominant view in contemporary psychology. Cognitive psychology is the study of memory processes: reasoning, attention, categorization, and problem solving. Until recently, its aim has been to study one species: Homo sapiens. Although cognitive scientists concede that the cognitive system of humans is a product of evolution, most reject the assumption that similar cognitive systems exist in other species of animals. Resistance to cognitive animal research comes partially from the pragmatic idea that any study must produce a result which has a direct and immediate application to humans. But the study of the behaviors of animals has produced psychological theories whose principles have enhanced our understanding of human behavior as much as have direct studies of humans. Philosophically, resistance comes from preconceived notions of human uniqueness. Yet many phenomena which have led to this claim have been demonstrated in a nonhuman animal. Such "human" behaviors as crossmodal perception, tool use, self-awareness, consciousness, mind, mental illness, murder, religion, culture, and insight have been demonstrated either in species of monkeys, pigeons, porpoises, or sea otter (see Gallup Jr. & Suarez, 1983 for a review.)

As a consequence of resistance to animal research (for pragmatic or philosophical reasons), cognitive psychology remains a somewhat egocentric discipline with an inevitable bias toward the human mind. Consequently, theories of human cognition have not been integrated with theories of animal learning. Comparative cognition resolves this failure of integration by accepting the assumption that there is some continuity of evolved mental processes across species.

## COMPARATIVE COGNITION: DEFINED AND QUALIFIED

Comparative cognition is the determination of any subsequent study of cognitive processes which are species-general. A species-general approach recognizes that basic cognitive processes exist across species. The rudiments of this approach are the same as those employed by human cognitive psychologists: Evidence for the presence of a cognitive process is acquired from a sample in a population, e.g., college students, and then, in order to determine the general applicability of the cognitive process, such evidence is sought from other populations, e.g., clinically or developmentally different populations. For example, Hasher and Zacks, (1979, 1984) first identified an automatic process of encoding frequency information by studying college students, and then demonstrated the generality of the process in populations of humans at different developmental stages and clinical states. The comparative cognition approach searches for general processes also, but the tested populations are different species, not clinically or developmentally different groups of the same species.

Comparative cognition is theoretically based on Darwin's (1871/1981) assumption that cognitive processes evolved. As he states, "The difference in mind between man and the higher animals, great as it is, certainly is one of degree and not of kind [p. 105]." The crux of this assumption is that differences in learning and thinking between species are quantitative, not qualitative.

Although its basis is evolutionary theory, one critical theoretical difference exists between the general processes approach and the Darwinian notion of an evolutionary scale of mind. The study of comparative cognition described here is not concerned with comparing the "intelligence" of species in order to find the intelligence quotient of each. The product of this study is not an evolutionary scale of cognitive processes. We agree with Hodos (1982) that we must abandon a unilinear hierarchical model for cognitive processes and the brain if we expect to make progress in the comparative study of cognitive phenomena and their neural substrates. Our strategy for studying cognitive processes is to identify those processes which are present in many species without quantifying them along some psychological scale.

There are two tasks for comparative cognition: (1) to determine the nature and limits of animal cognition; and (2) to compare cognitive processes of different species. We are concerned with the first task, determining the nature of animal cognition. This study is well underway. For example, Herrnstein and deVilliers (1980) found that pigeons can classify novel items into previously learned categories; and Sands, Lincoln, and Wright (1982) discovered that monkeys could separate a group of items into categories. In Chapter 7, Kendrick and Rilling cite evidence which suggests that a similar division of active and inactive memory exists in pigeons and in people. Wright, Santiago, Sands, and Urcuioli (1984) have found comparable serial position curves in data from pigeons, monkeys, and people in a serial probe recognition task.

Riley (1984) has found support for a limited-capacity hypothesis and selective attention in pigeons. Moreover, directed forgetting has been demonstrated in people (Bjork, 1972), rats (Grant, 1982), monkeys (Roberts, Mazmanian, & Kraemer, 1984), and in pigeons (Maki & Hegvik, 1980; for a review, see Rilling, Kendrick, & Stonebraker, 1984).

We assume there are no basic cognitive processes—such as internal representation of the environment, pattern recognition, categorization, image formation, or reasoning—that are uniquely human, or unique to any one species. However, the degree to which cognitive processes are developed and integrated makes them uniquely functional in different species. Thus the second task of comparative cognition is to compare cognitive processes of different species. This study is not antagonistic to a general-processes approach, as long as it does not reject the body of literature which demonstrates that species share basic common cognitive processes. Scientists who ignore this literature have been unsuccessful at demonstrating that species-specific differences cancel out or cover up the effects of general processes.

An example of an attempt to refute the existence of general processes is found in the taste-aversion learning controversy. Although researchers found that manipulation of certain variables yielded different outcomes in different species, Logue's (1979) evaluation of such studies showed that no new principles were needed to explain the varying results within these studies of taste-aversion learning. She concluded that common processes, such as one-trial learning, specificity of the conditioned stimulus to the unconditioned stimulus, and blocking existed in studies of each species. In their review, Domjan and Galef (1983) came to the same conclusion. General processes prevailed in taste-aversion learning, although the effects of certain variable manipulations changed for different species.

The controversy about biological constraints on cognition is reminiscent of the controversy about taste-aversion learning. For example, Terrace (1984) advocates a biological constraints approach to animal cognition. He assumes that there are qualitative differences between human and animal representations, and he cautions against extrapolation from animal to human cognitive processes. Kamil (1984) concludes that the study of animal cognition is of little interest per se because of the failure of the general process approach to animal learning. This chapter is written as an alternative to the metatheories advocated by Kamil and Terrace. Unfortunately, the data base of animal cognition research is not large enough to allow the definitive review to be written. We have found two examples which support a general processes approach to animal cognition even when constraints on cognition might be expected.

## General Processes Versus Biological Constraints

Field experiments with species-specific behaviors are precisely the conditions in which biological constraints on cognition are held to be most likely to occur.

Birds in the southwestern US exhibit varying degrees of anatomical specialization depending upon the degree to which they cache food. Clark's nutcracker shows the most specialization and has evolved a sublingual pouch under the tongue in which seeds are carried (see Shettleworth, 1983 for detail). Given this anatomical specialization, we may ask if evolution has produced cognitive structures as an adaptive specialization enabling this bird to recover hoarded food. From the standpoint of a general process approach to animal cognition, we may ask if this cognitive skill of recovering cached food requires unique cognitive processes or if the differences between species are quantitative thereby supporting a general process approach.

Each autumn, the Clark's nutcracker may cache up to 33,000 seeds. This hoarding is clearly a species-specific behavior and has been studied as such by ethologists (Vander Wall, 1982). The seeds are recovered during the winter and they constitute between 80–100% of the bird's diet (Balda & Turek, 1984). Up to 11 months may elapse between caching the seeds and subsequent recovery. Evidence that memory is involved in recovering the food comes from experiments by Vander Wall and Balda and Turek in which stones on a sandy floor on which a nutcracker cached seeds were moved from their original locations. The birds consistently made errors in the direction toward which the stones were moved. In other words, nutcrackers appear to remember the location of the stones, not the location of the seeds.

Balda and Turek's (1984) discussion of memory in birds demonstrates no need for processes unique to the nutcracker. The stimuli in the environment which were represented in the animal's long-term memory appear to be a combination of small surface cues, such as rocks and logs, which serve as general cues to the location of the cache site. The search is controlled by a cognitive map stored in long-term memory.

Shettleworth and Krebs (1982) have carried out a series of experiments on food hoarding with marsh tits. As with the nutcracker, the marsh tit uses memory for recovering the hoarded food. A recency effect, first recovery of seeds stored last, was found two hours after storing several batches of seeds. Shettleworth and Krebs found that the marsh tits were extremely good at this cognitive skill, recovering the first five seeds by inspecting 10 holes on average, so about half of their early recovery attempts were successful. Excellent spatial memory is not unique to birds: performance of rats in spatial memory tasks is well known (Olton, 1978; Roberts & Dale, 1981; see Roberts, 1984 for a review). Shettleworth and Krebs conclude that spatial memory is a quantitative adaptive specialization, not a qualitative specialization unique to food-hoarding animals. A general processes approach provides a good account of spatial memory data.

A comparison of the cognitive processes of different species is also a research strategy in which biological constraints on cognitive processes are held most likely to occur. Rumbaugh (1971) and Rumbaugh and Pate (1984) have followed the strategy of searching for qualitative differences in cognition by comparing the performance of various primate species (as well as a clinical sample of retarded human adolescents) on learning set tasks. The tasks consisted of a long series of two-choice visual-discrimination problems. An index of transfer on reversal test trials was the datum of interest. The experiment was designed to determine the extent to which a species employed associative as opposed to mediational mechanisms of learning. Clear quantitative differences among species in the amount of transfer emerged: The primitive prosimians and the talapoin monkeys showed strong negative transfer, while the apes showed positive transfer. Furthermore the results fit relatively neatly on the evolutionary scale with increases in brain complexity associated with a shift from negative to positive transfer.

Rumbaugh's (1971) original view, and the impetus for his research, was that the principles necessary to explain cognition in more highly evolved forms are different from the principles necessary to explain cognition in less highly evolved forms. However, the data that Rumbaugh and Pate (1984) provide can be interpreted in terms of a quantitative weighting for each species of four concepts from general process learning theory and general process cognitive findings: (1) excitation of response strength to the reinforced stimulus; (2) inhibition of response strength to the extinguished stimulus; (3) the tendency to select a novel stimulus; and (4) an ability to respond to the relationships among stimuli as opposed to responding to specific stimuli. No cognitive process unique to one of the species studied provides a better interpretation of the data than does the assumption of continuity of cognitive processing among species. Moreover, relational associations are not unique to primates. Wright, Santiago, Sands, and Urecioli (1984) found that relational associations develop prior to item-specific associations in pigeons, monkeys, and humans. Rumbaugh and Pate's and Wright and Wright et al.'s studies provide good support for the notion that general cognitive processes exist across species.

## COMPARATIVE COGNITION: OPERATIONALLY DEFINED

The study of comparative cognition applies behaviorist methodology to the study of cognitive processes in animals. The data base of comparative cognition is accumulated by presenting stimuli to animals and recording their responses. By conducting many studies and analyzing behaviors induced by experimental manipulations, cognitive processes are identified by inference. Cognitive processes are identified when they are required to explain the behavioral results. Crucial components of this analysis are reviewed: independent variables, dependent variables, and the analysis of converging operations which leads to the identification of cognitive processes.

## Independent Variables

The independent variables which are used to study cognitive issues are more restricted in the study of animal cognition than they are in human cognition. Semantic or language manipulations are impossible, unless of course symbols and "language" are trained, and even then, the validity of studying the trained processing of language and symbols may be questioned. More basic processes which do not necessarily involve verbal skills and which can be tested without extensive practice (i.e., categorization, memory capacities, reasoning, and numerosity) can be studied. Research on animal cognition is closely related to previous operant research on stimulus control (Rilling, 1977) and animal psychophysics (Blough & Blough, 1977). Within an experimental situation, environmental events (usually visual or auditory stimuli) are changed (usually temporally or spatially). The difference between operant and cognitive experiments is that different independent variables serve paramount roles. The most important variable in operant conditioning is the schedule of reinforcement, punishment, or extinction. The emphasis is on how environmental events which follow responding affect responding. Of primary interest in cognitive work is the time in which a stimulus is on or off, or the time between stimulus presentation and a cue to respond. The emphasis here is on whether responding is affected when sensory processes are restricted, or when memory is taxed. Both operant and animal cognitive experiments manipulate environmental events to affect changes in responding. Their manipulations are controlled by what theoretical issues are being investigated: prediction and control of behavior, or memorial and cognitive processes involved in behavior.

## Dependent Variables

The cognitive processes of people and of other animals are not observable directly. The problem for cognitive science is the selection of appropriate dependent variables, i.e. directly observable behavior, for the study of cognitive phenomena. Since many animal cognitive scientists studied operant conditioning, it is not surprising that the typical dependent variable collected is either rate of response or percent correct response. Again there is a specific strategy for analyzing these operant variables in cognitive experiments. The critical variable in cognitive study is transfer to novel stimuli. If percent correct response (or rate of response) remains high in novel situations, we then have evidence of the use of a cognitive process. Suppose that animals are trained to respond in a certain way. We hypothesize that they are using a cognitive process in order to perform well. To verify our hypothesis, those animals must respond equally well in novel situations in which learning of discrete stimulus-response (S-R) associations is not applicable, but employing the cognitive process is applicable. This strict analysis of percent correct response (or rate of response) makes the identification of cognitive processes in animals extremely difficult. Changing environmental situations invariably affects responding, but responding must be maintained in novel situations in order to demonstrate the presence of a cognitive process. Still, cognitive processes have been identified by this method of analysis in various species of birds, rats, monkeys, and in people.

Reaction time, or the time taken to respond, is commonly collected in human cognitive experiments as an index of cognitive processes at work. Whether reaction time is a viable dependent variable in the study of animal cognition remains to be determined, since reaction time is not yet commonly collected in most animal studies. Blough (1984) collected reaction times of pigeons in a pattern-recognition task and found that reaction times increased as the number of items which required scanning increased. There is some indication, at least in this study, that reaction times reflected cognitive processes at work. According to human cognitive psychologists, reaction time is a meaningful variable for measuring cognitive processes only if subjects respond as fast as they possibly can. In experiments with animals, subjects must be forced to respond within a certain amount of time in order for reaction times to be meaningful. This might be achieved by manipulating contingencies of reinforcement to include a time constraint on responding. Speed-accuracy trade-offs are sure to incur. To establish reaction time as a viable dependent variable for animal cognitive work requires extensive investigation.

## Converging Operations

Validation using a converging set of operations (Garner, 1981) is required to invoke an explanation of a process which cannot be directly observed. Convergent validation occurs when a theoretical process is supported by different experiments. These experiments may vary with respect to the task employed, the subject pool used, and the data collected, but the results of these studies converge on, or point to, a single theoretical construct which best explains those results. In order to establish evidence to support a cognitive interpretation, the standard procedure in comparative cognition is convergent validation. To illustrate the strength of this methodology, consider how rehearsal has been established by the method of convergent validation as a cognitive process found in animals.

## THE GENERAL PROCESS APPROACH APPLIED: REHEARSAL

Consider asking the question, "Do animals rehearse?" We define rehearsal as the maintenance of a representation in short-term memory after the removal of a stimulus to be remembered. Operationally, a rehearsal process takes place after a stimulus presentation has been terminated, and before memory is tested. In humans, rehearsal is effected when: (1) directed forgetting occurs; (2) a distractor task is presented during the time in which rehearsal occurs; (3) the number of items to be remembered is increased, thus taxing the rehearsal process; or (4) the stimulus to be remembered is novel or surprising, thus activating rehearsal. Each of these affectors of rehearsal have been demonstrated in animals.

Directed forgetting is a cognitive phenomenon in which subjects are unable to recall past events for which they were cued to "forget." The effect on rehearsal is that the "forget" cue inactivates a rehearsal process, and since the event was not rehearsed, it is not stored in memory and is forgotten. The study of directed forgetting in animals demonstrates this type of stimulus control of a "forget" cue on the rehearsal process (see the review by Rilling, Kendrick, & Stonebraker, 1984). The typical procedure involves a matching-to-sample task in which instructional cues (meaning "remember" and "forget") are presented in the delay interval between sample stimulus presentation and the test for retention. There is no test for retention following the "forget" cue in this task, and there is always a test for retention following the "remember" cue. To demonstrate directed forgetting, a "forget" cue is followed by a test for retention in violation of previous training. The result is that memory for the sample stimulus in this type of trial is relatively poor, which indicates that the "forget" cue induced forgetting. The phenomenon indicates that a rehearsal process maintains a representation of the stimulus to be remembered, and this rehearsal process can be inhibited by a "forget" cue. Similarities in the directed forgetting phenomenon between humans and other animals (Grant, in press; Kendrick and Rilling, 1984; Maki, 1981) imply that a basic rehearsal process is being similarly affected across species.

The effect on rehearsal of a distractor task is that memory is worse when the task interferes with the rehearsal process. Kendrick and Rilling (1984) trained pigeons on a successive delayed matching-to-sample task on one side key, and presented stimuli in distractor tasks on the other side key during the retention interval (the time between sample stimulus presentation and the test). They found that schedules of reinforcement, which distracted subjects from the successive matching-to-sample task by disrupting pre-established delay-interval behavior, produced forgetting in the matching-to-sample task. Pigeons were rehearsing the stimuli during the delay interval in the successive matching-to-sample task. If rehearsal was interrupted, memory for stimuli was worse.

As the number of items to be remembered is increased, there is a tendency for humans to remember those items which occurred at the beginning of the presentation and those at the end, i.e. those items rehearsed most, and those items most recent in short-term memory. This primacy-recency effect produces the familiar serial position curve. Sands and Wright (1980) employed photographs of stimuli as a list of items in a serial probe recognition task which was developed to study primacy and recency effects in animal memory. In subsequent study (Wright et al., 1984) they found that rhesus monkeys and pigeons produce a serial position curve very similar in form to that obtained by humans. Monkeys, pigeons, and the human they tested also showed the same systematic changes in the form of the serial position function with changes in the delay preceding test stimulus presentation. Each species' ability to remember items was affected by changes in the retention interval during which rehearsal occurs.

Novel or surprising stimuli are best remembered because novel stimuli draw attention and may require active rehearsal. Wagner (1978) studied memory for surprising stimuli (i.e. a previously conditioned negative stimulus with an unconditioned stimulus) and expected stimuli (i.e. a conditioned positive stimulus with an unconditioned stimulus) in pigeons. He found that a surprising unconditioned stimulus caused more interference with the acquisition of a conditioned response than did an expected unconditioned stimulus. He concluded that surprising unconditioned stimuli might require active rehearsal in the pigeon. This challenged trace-decay theory, which assumed that the representation of a stimulus passively decayed after stimulus termination.

Maki (1981) extended the research on surprising versus expected stimuli to the delayed matching-to-sample paradigm. In his experiment, pigeons' responding was more effected by surprising sample-reinforcement pairs than by expected sample-reinforcement pairs, indicating that pigeons better remembered surprising events than expected events. Grant, Brewster, and Stierhoff (1983) confirmed Maki's finding that accuracy was higher on surprising probes than on expected probes. They also showed that the surprising-expected effect was influenced by changes in the delay interval during which rehearsal might take place. For surprising probes, lengthening the delay did not effect the high level of accurate responding. For expected probes, lengthening the delay retarded performance rapidly. Here were clear examples of rehearsal retarding forgetting and of active rehearsal of novel stimuli facilitating remembering in pigeons.

Four characteristics of rehearsal in people are: (1) rehearsal is inactivated by directed forgetting; (2) it is prevented by presenting a distractor task during the delay interval; (3) it is effected by the number of items to be remembered and the amount of time during which they are rehearsed; and (4) surprising events induce active rehearsal and are remembered better than expected events. Each of these characteristics of rehearsal has been found in research on animals. Although the species studied and methodology used differ from study to study, the data from this converging set of operations clearly point to a cognitive interpretation based on a rehearsal process.

## ILLUSTRATIONS OF A GENERAL PROCESSES
## APPROACH

Is the general processes approach accepted and employed by researchers in the field of comparative cognition? Do animal-cognitive scientists assume that basic cognitive processes exist across species? In order to find out, we carried out an informal survey of animal cognition researchers in search of parallels between human and animal cognition. The two topics we review here are language processes and visual-information processes.

### Language Processes in Animals

The topic of language processes in animals is the most controversial area of animal cognition. However, developmental psychology provides a foundation for the study of communicative skills and symbol learning in animals. Each human infant goes through a series of development phases involved in language acquisition, from self-oriented to intentional communication (Savage-Rumbaugh, 1984). A general process approach to this topic, illustrated by the work of Savage-Rumbaugh, demonstrates that the process of symbol acquisition by chimpanzees is fundamentally similar to that of human children. Naming with reference has been demonstrated in chimpanzees in the following ways: (1) choice of an item to be named from a set of possible foods; (2) naming the object by pressing the symbol for the object on a keyboard; and (3) correct selection of the food from a set of possible foods. Concepts like "intentionality" are operationally defined. These results have been obtained under conditions in which possible cues from the experimenter have been eliminated, a problem with previous work that was noted by Terrace (Terrace, Petitto, Sanders, & Bever, 1979). Symbol learning has been demonstrated in chimpanzees (Premack, 1983) as well as in pigeons (Zentall & Hogan, 1978) and in dolphins (Herman, Richards, & Wolz, 1984).

The study of nonhuman artificial language in a variety of species is plagued with methodological difficulties. However, the work of Savage-Rumbaugh (1984) demonstrates the viability of a general process approach to the study of communication among animals. By following a general process approach to language—i.e. by investigating communication among species and symbol learning in animals—it is possible to identify basic functions of communication and capacities for symbol learning. Additionally, it might also be possible to determine the characteristics of language which are uniquely human from this analysis.

### Visual Information Processing in Animals

Visual information processing (see Pinker, 1984) is a well-developed area of cognitive science that readily lends itself to the study of cognitive processes in animals. From the standpoint of our general process approach to animal cognition, we may ask if basic visual information processing skills in pigeons or other animals are similar to those basic processes in people. For example, similarity in visual information processing between people and pigeons was found by Blough (1982). He presented letters of the English alphabet to pigeons in a simple discrimination task and found similarity patterns in pigeons that were well correlated ($r = 0.68$) with comparable data from human subjects. He found cross-species generality in the processing of letters, which supports a general process approach to visual information processing.

Characteristics of visual search as a visual information process have been investigated in humans. Most notably, Triesman and Gelade (1980) demonstrated that two different search processes can be activated: serial and parallel search. Serial search is activated if the subject is looking for a target stimulus in a group of heterogenous or unrelated stimuli (i.e., looking for an "A" in a screen full of alphabet characters). The behavioral result of serial search is that the time it takes to complete the process, finding the "A" in this case, is lengthened if the number of stimuli to be searched is increased. The other process, parallel search, is activated if the subject is looking for a target stimulus which is unique and unrelated to the distractor stimuli (i.e., looking for a "1" on a screen full of alphabet characters). The time taken to parallel search is not effected by the number of stimuli present.

Patricia Blough (1984) has developed a procedure for investigating visual search in pigeons in which data parallel the human findings. The target stimuli were letters of the alphabet presented on a video monitor controlled by a microprocessor. The pigeon's task was a simple discrimination of pecking the one of three keys behind which a single target letter was located. The independent variable was the number of distractor letters. The results showed that as the size of the set in which the pigeon had to search for the target increased from three to nine letters, the reaction time increased. In this experiment, reaction time was more sensitive to the size of the display than was accuracy, a clear demonstration that chronometric methods, so popular in human cognitive psychology, can be successfully employed with pigeons. Additionally, Blough found that if distractor letters were the same, increasing the number of them had a minimal effect on reaction time. She provided behavioral evidence that pigeons serial and parallel search just as humans do.

Although the roots of picture memory can be traced back into the nineteenth century, picture processing as a visual information process has received renewed importance. In particular, two characteristics demonstrated in human memory of pictures are a large capacity and a long retention. The capacity of people's long-term memory for pictures is measured by showing them a set of pictures and providing a few seconds to view each. In the test for long-term memory, people are shown a mixture of old and new pictures and asked to identify whether the picture is old or new. The general finding is that subjects' accuracy for a set of several thousand photographs is over 90% correct when the recognition test is given immediately after presentation of the original pictures (Spoehr & Lehmkuhle, 1982). This finding led theorists to conclude that the long-term capacity of the human memory system for pictures is essentially unlimited.

Vaughan and Greene (1984) have developed a procedure using picture memory to measure the long-term memory capacity of the pigeon. Rather than asking the pigeon if the picture was old or new, Vaughan and Greene asked the pigeon to remember if the picture was associated with reinforcement or non-reinforcement. The stimuli were photographic slides randomly assigned to a positive or negative set independently of any category within the natural structure of the stimuli themselves. Although acquisition of the discrimination required about 1,000 sessions of training, the results demonstrated that pigeons can discriminate at least several hundred different stimuli. As with humans, the pigeon has a very large capacity for picture recognition.

A second characteristic of human picture recognition is long-term retention. This characteristic is shared by the pigeon as well. Vaughan and Greene (1984) found that following a retention interval of 490 days, only a slight decrement in retention was obtained. After more than 2 years, the birds continued to exhibit excellent long-term retention. Pigeons and people share two characteristics of picture memory: a large long-term memory capacity and excellent long-term retention.

Organization of the stimuli in the picture is a major determinant of picture recognition in people (Spoehr & Lehmkuhle, 1982). One of the gestalt principles of organization which has re-emerged from the renewed interest in visual information processing is context effects: A form is more readily discriminated than an unrelated set of stimuli even though the discriminative information in the two sets of stimuli is identical. P. M. Blough (1984) found that the acquisition of a simple discrimination in pigeons between two line angles was facilitated when the lines were embedded in a pattern containing a right angle. Blough provided evidence that the gestalt principles of context effects may be applicable to pigeons as well as people. Visual information processing provides a good framework for the study of cognitive processes in pigeons. The initial findings support a general process approach.

## CONCLUSION

The general processes approach to comparative cognition accepts Darwin's (1871/1981) thesis that there is some evolutionary continuity of mind across species. The general processes approach rejects a hierarchical study of species' cognitive processes, and instead concerns itself with the identification of basic cognitive processes which exist across species. A crucial assumption is that no basic cognitive processes are unique to any species. Although the development and integration of cognitive processes within species may provide functional uniqueness, the general processes approach searches for similarities in cognitive processing among species and seeks to identify those cognitive processes which generalize across genetic differences among species. Examples which were provided to illustrate this approach include studies on rehearsal, language or symbol learning, and visual information processing.

The explosive development of cognitive science, both as a methodology and as a theoretical orientation, provides a framework for the study of cognitive phenomena in animals. The cognitive system of any animal (humans included) is a system with a bias to process information selectively, to store certain relationships associatively, and to ignore other relationships. To determine the operating characteristics of a cognitive system, i.e., to analyze the bias, an extensive experimental analysis of responses to changes in environmental and physical parameters of stimuli is required. In order to determine the limits and plasticity of the cognitive system under examination, a necessary step is to present novel stimuli not encountered in the animal's environment. The results of such laboratory work will measure the ability and capacity of the cognitive system to track environmental change, and the generality of existing capabilities across species as well. Cognitive phenomena once considered unique to humans have been obtained in laboratory animals. From this research effort, a picture of the general nature of basic cognitive processes which exist despite environmental change and genetic differences is beginning to emerge.

# REFERENCES

Balda, R. P., & Turek, R. J. The cache-recovery system as an example of memory capabilities in Clark's nutcracker. In H. L. Roitblat, T. G. Bever & H. S. Terrace (Eds.), *Animal cognition.* Hillsdale, NJ: Lawrence Erlbaum Associates, 1984.

Bjork, R. A. Theoretical implications of directed forgetting. In A. W. Melton, & E. Martin (Eds.), *Coding processes in human memory.* Washington DC: Winston, 1972.

Blough, D. S. Reaction times of pigeons on a wavelength discrimination task. *Journal of the Experimental Analysis of Behavior,* 1978 *30,* 163–167.

Blough, D. S. Pigeon perception of letters of the alphabet. *Science,* 1982, *218,* 397–398.

Blough, D. S. Form recognition in pigeons. In H. L. Roitblat, T. G. Bever, & H. S. Terrace (Eds.), *Animal cognition.* Hillsdale, NJ: Lawrence Erlbaum Associates, 1984.

Blough, P. M. Visual search in pigeons: Effect of memory set size and display variables. *Perception and Psychophysics,* 1984, *35,* 344–352.

Blough, D. & Blough, P. Animal Psychophysics Fn. W. K. Honig & J. E. R. Staddon (Eds.), *Handbook of operant behavior.* Englewood Cliffs, NJ: Prentice-Hall, 1977.

Darwin, C. *The descent of man, and selection in relation to sex.* Princeton, NJ: Princeton University Press, 1981. (Originally published, 1871).

Domjan, M., Galef, B. G. Biological constraints on instrumental classical conditioning: Retrospect and prospect. *Animal Learning and Behavior,* 1983, *11,* 151–161.

Gallup, G. G., Jr., & Suarez, S. D. Overcoming our resistance to animal research: Man in comparative perspective. In D. W. Rajecki (Ed.), *Comparing behavior: Studying man studying animals.* Hillsdale, NJ: Lawrence Erlbaum Associates, 1983.

Garner, W. R. The analysis of unanalyzed perceptions. In M. Kubovy & J. R. Pomerantz (Eds.), *Perceptual organization*. Hillsdale, NJ: Lawrence Erlbaum Associates, 1981.

Grant, D. S. Stimulus control of information processing in rat short-term memory. *Journal of Experimental Psychology: Animal Behavior Processes*, 1982, *8*, 154–164.

Grant, D. S. Directed forgetting and intertrial interference in pigeon delayed matching. *Canadian Journal of Psychology*, 1984, *38*, 166–177.

Grant, D. S., Brewster, R. G., & Stierhoff, K. A. "Surprisingness" and short-term retention in pigeons. *Journal of Experimental Psychology: Animal Behavior Processes*, 1983, *9*, 63–79.

Hasher, L., & Zacks, R. T. Automatic and effortful processes in memory. *Journal of Experimental Psychology: General*, 1979, *108*(3), 356–388.

Hasher, L. & Zacks, R. T. Automatic processing of fundamental information: The case of frequency of occurrence. *American Psychologist*, 1984, *39*, 1372–1388.

Herman, L. M., Richards, D. G., & Wolz, J. P. Comprehension of sentences by bottlenosed dolphins. *Cognition*, 1984, *16*, 1–90.

Herrnstein, R. J., & deVilliers, P. A. Fish as a natural category for people and pigeons. In G. H. Bower (Ed.), *The psychology of learning and motivation, vol. 14*. New York: Academic Press, 1980.

Hodos, W. Some perspectives in the evolution of intelligence and the brain. In D. R. Griffin (Ed.), *Animal mind–human mind*. New York: Springer-Verlag, 1982.

Kamil, A. C. Adaptation and cognition: Knowing what comes naturally. In H. L. Roitblat, T. G. Bever, & H. S. Terrace (Eds.), *Animal cognition*. Hillsdale, NJ: Lawrence Erlbaum Associates, 1984.

Kendrick, D. F., & Rilling, M. R. AIM: a theory of active and inactive memory. In D. F. Kendrick, M. R. Rilling, & M. R. Denny (Eds.), *Theories of animal memory*. Hillsdale, NJ: Lawrence Erlbaum Associates, 1984.

Kendrick, D. F., & Rilling, M. R. The role of distractor tasks in retroactive inhibition in pigeon short term memory. *Animal learning and behavior*, *12*, 391–401.

Logue, A. W. Taste aversion and the generality of the laws of learning, *Psychological Bulletin*, 1979, *86*(2), 276–296.

Maki, W. S. Directed forgetting in animals. In N. E. Spear & R. R. Miller (Eds.), *Information processing in animals: Memory mechanisms*. Hillsdale, NJ: Lawrence Erlbaum Associates, 1981.

Maki, W. S., & Hegvik, D. K. Directed forgetting in pigeons. *Animal Learning and Behavior*, 1980, *8*, 567–574.

Olton, D. S. Characteristics of spatial memory. In S. H. Hulse, H. Fowler, & W. K. Honig (Eds.), *Cognitive processes in animal behavior*. Hillsdale, NJ: Lawrence Erlbaum Associates, 1978.

Pinker, S. Visual cognition. *Cognition*, 1984, *18*, 1–63.

Premack, D. The codes of man and beasts. *The Behavioral and Brain Sciences*, 1983, *6*(1), 125–169.

Riley, D. A. Do pigeons decompose stimulus compounds? In H. L. Roitblat, T. G. Bever, & H. S. Terrace (Eds.), *Animal cognition*. Hillsdale, NJ: Lawrence Erlbaum Associates, 1984.

Rilling, M. R. Stimulus control and inhibitory processes. In W. K. Honig & J. E. R. Staddon (Eds.), *Handbook of operant behavior*. Englewood Cliffs, NJ: Prentice-Hall, 1977.

Rilling, M, Kendrick, D. F., & Stonebraker, T. B. Directed forgetting in context. In G. H. Bower (Ed.), *The psychology of learning and motivation, Vol 18*. New York: Academic Press, 1984.

Roberts, W. A. Some issues in animal spatial memory. In H. L. Roitblat, T. G. Bever, & H. S. Terrace (Eds.), *Animal cognition*. Hillsdale, NJ: Lawrence Erlbaum Associates, 1984.

Roberts, W. A., & Dale, R. H. I. Rembrance of places lasts: Proactive inhibition and patterns of choice in rat spatial memory. *Learning and Motivation*, 1981, *12*, 262–268.

Roberts, W. A., Mazmanian, D. S., & Kraemer, J. Directed forgetting in monkeys. *Animal Learning and Behavior*, 1984, *12*, 29–40.

Rumbaugh, D. M. Evidence of qualitative differences in learning processes among primates. *Journal of Comparative and Physiological Psychology,* 1971, *76,* 250–255.

Rumbaugh, D. M., & Pate, J. L. The evolution of cognition in primates: A comparative perspective. In H. L. Roitblat, T. G. Bever, & H. S. Terrace (Eds.), *Animal cognition.* Hillsdale, NJ: Lawrence Erlbaum Associates, 1984.

Sands, S. F., Lincoln, C. E., & Wright, A. A. Pictorial similarity judgments and the organization of visual memory in the rhesus monkey. *Journal of Experimental Psychology: General,* 1982, *111*(4), 369–389.

Sands, S. F., & Wright, A. A. Primate memory: Retention of serial list items by a rhesus monkey. *Science,* 1980, *209,* 938–939.

Savage-Rumbaugh, E. S. Acquisition of functional symbol usage in apes and children. In H. L. Roitblat, T. G. Bever, & H. S. Terrace (Eds.), *Animal cognition.* Hillsdale, NJ: Lawrence Erlbaum Associates, 1984.

Shettleworth, S. Memory in food-hoarding birds. *Scientific American,* 1983, *248,* 102–110.

Shettleworth, S. J., & Krebs, J. R. How marsh tits find their hoards: The roles of site preference and spatial memory. *Journal of Experimental Psychology: Animal Behavior Processes,* 1982, *8,* 354–375.

Spoehr, K. T., & Lehmulke, S. W. *Visual information processing,* San Francisco, CA: W. H. Freeman, 1982.

Terrace, H. S. Animal cognition. In H. L. Roitblat, T. G. Bever, & H. S. Terrace (Eds.), *Animal cognition.* Hillsdale, NJ: Lawrence Erlbaum Associates, 1984.

Terrace, H. S., Petitto, L. A., Sanders, R. J., & Bever, T. G. Can an ape create a sentence? *Science,* 1979, *206,* 891–900.

Triesman, A. M., & Gelade, G. A feature integration theory of attention, *Cognitive Psychology,* 1980, 97–136.

Vander Wall, S. B. An experimental analysis of cache recovery in Clark's nutcracker. *Animal Behavior,* 1982, *30,* 84–94.

Vaughan, W., Jr., & Greene, S. L. Pigeon visual memory capacity. *Journal of Experimental Psychology: Animal Behavior Processes,* 1984, *10*(2), 256–271.

Wagner, A. R. Expectancies and the priming of short term memory. In S. H. Hulse, H. Fowler, & W. K. Honig (Eds.), *Cognitive processes in animal behavior.* Hillsdale, NJ: Lawrence Erlbaum Associates, 1978.

Wright, A. A., Santiago, H. C., Sands, S. F., & Urcuioli, P. J. Pigeon and monkey serial probe recognition: Acquisition, strategies, and serial position effects. In H. L. Roitblat, T. G. Beuer, & H. S. Terrace (Eds.), *Animal cognition.* Hillsdale, NJ: Lawrence Erlbaum Associates, 1984.

Wright, A. A., Santiago, H. C., Urcuioli, P. J., & Sands, S. F. Monkey and pigeon acquisition of same/different concept using picturial stimuli. In M. L. Commons, & R. J. Herrnstein (Eds.), *Quantitative Analysis of Behavior, Vol. 4,* 1984.

Zentall, T. R., & Hogan, E. Same/different concept learning in the pigeon: The effect of negative instances and prior adaptation to transfer stimuli. *Journal of the Experimental Analysis of Behavior,* 1978, *20,* 177–186.

# 3 "Retention" of S-R in the Midst of the Cognitive Invasion

M. Ray Denny
*Michigan State University*

The major premise of this chapter is that explicit continuity between the more traditional Stimulus-Response (S-R) framework and recent cognitive formulations can be readily established and exploited. The bridge, or at least its foundation, lies in appropriately defining stimulus (S) and response (R) (Denny, 1966, 1967, 1971; Denny & Ratner, 1970).

There are at least three advantages to conceptualizing the subject matter of psychology in appropriately defined S-R terms. First, this approach provides us with a consistent system that relates equally well to all behavior. Psychology now consists of disparate, fragmented systems that discourage integration; and cognitive psychology intrinsically lacks coherence (Schnaitter, 1983). With a revised S-R framework, the possibility for much-needed integration would be enhanced. Second, the various systems currently tend either to ignore each other or squabble with each other. This wasteful practice means that the cognitive theorists tend to eschew the bulk of the research findings that were obtained earlier by S-R psychologists and that traditional S-R psychologists tend to retain an incomplete picture of behavior. Since neither camp has a corner on the whole "truth," this is a self-defeating state of affairs. Third, a cognitive position, with its emphasis on mind or mental events, runs the strong risk of promoting and prolonging dualistic and mentalistic views of behavior.

The breath of fresh air provided by the S-R, behavioristic tradition, if overthrown, could set back an objective, scientific view of behavior many years. Even though many animal cognitivists steer clear of mentalism (Wasserman, 1983), this practice does not guarantee an objective behavioral interpretation by others and may even encourage the opposite because of the nature of the terminology. We simply cannot afford a regression to dualism before we've even shed its remnants. The best of all worlds would be if the general public increased its use of behavioral terms, as has happened somewhat with the terminology of other scientific disciplines (Pear, 1983). Then the distinction between cognitive and behavioral terminologies might disappear.

Finally, cognitive terminology is the language of everyday speech, and because of this, despite its advantage for communicating with the man in the street, is metaphorical and imprecise (Guilford, 1982). It also includes misconceptions that were built into the language over its long span of development. For all of these interrelated reasons, the innumerable evolutionary and experiential variables that determine behavior, of which conscious humans are generally unaware, can be seriously shortchanged within a strictly cognitive framework.

## DEFINING STIMULUS AND RESPONSE

S-R formulations in general have been inappropriately constituted. Traditionally, both the concept of "stimulus" and the concept of "response" have been viewed essentially as observables, close to the "thing" or empirical level like the low-order concepts of table and chair (Lewis, 1963) instead of like length or velocity. Stimuli have been typically identified as external objects; and responses, as response occurrences or instances of behavior, have not been sufficiently distinguished from response classes. But the S and R in S-R generalizations are high-order abstractions, as is true of all concepts that are related to each other in empirical laws, and should be recognized as such. In the present formulation, response classes obviate the need for intervening variables or hypothetical concepts and simply dictate the use of a detailed analysis of objective data—the stimulus situation and response occurrences.

Even though S and R are usually viewed conceptually as quite low-order abstractions, in actual practice, the concepts are tacitly used appropriately. That is, when a particular stimulus is being specified in the *Procedure and Apparatus* section of an experimental report, all of the referents for this concept, either implicitly or explicitly, are dealt with: (1) the class of organism to which the reputed stimulus is applied is specified; (2) the "stimulus" normally is perceived or responded to by intact members of this class; and (3) its channel of administration is by means of appropriate sense organs or the like, rather than directly to the muscle, nerve, or motor neuron, in which case we would appropriately label the effect innervation. A stimulus cannot be equated with an object or event; it is an inference based on the three aforementioned types of referents, as well as on some object or event, and is therefore quite abstract.

Throughout a research report, references to a response class predominate over those to response occurrences. Rate of key pecking, median latency of bar pressing, mean amplitude of panel push, percentage of correct turns, mean number of errors, or the like are the typical ways of referring to response. This mathematization of response occurrences clearly denotes that we are fundamentally concerned with many similar response occurrences that constitute a distinguishable class of response. And some of the response class labels listed above nicely point up the practice of using the stimulus situation in which the responding occurs to help infer the appropriate classification. *Key, bar,* and *panel* as critical parts of the stimulus situation help provide a label for the appropriate response class. Moreover, common terms for various response classes—such as approach, avoidance, escape, fear, hostility, seeing, hearing, cooperativeness, and decision making—are usually sprinkled throughout the *Introduction and Discussion*. Obviously, it is response classes that interest us; and these classes are not directly observable but are inferred from certain types of referents.

Both stimulus and response as they enter into S-R laws are inferences. Such inferences, it should be stressed, are based on specifiable observables, typically on a set of several referents or definiens. Thus all response classes, no matter what their nature, have the same inferential status. A right turn by a rat in a T-maze, when properly classified, is just as much an inference as a -thought, percept, or image when so classified. Psychologists seem to be more confident about appropriately classifying explicit response occurrences than about classifying implicit events; but experience, as we shall see, may not justify that assumption.

Over 30 years ago, largely at the inspiration of graduate students who strongly believed in their philosophical analysis of psychology as a science (H. M. Adelman, R. A. Behan, R. H. Davis, J. L. Maatsch, and O. A. Smith), we attempted to formulate generic definitions of stimulus (S) and response class (R) that accurately reflected how psychologists were really using these concepts. These were not definitions of a particular stimulus or a particular response but guidelines for defining any S or any R. When these guidelines are spelled out, they have important implications.

## Definition of Stimulus (S)

As a concept, S is inferred from four referents. Briefly stated, S is defined (a) when some object or event (including behavioral events), (b) acting through the afferent neural pathways, is (c) potentially capable of eliciting an R, in (d) a particular class of organisms. *Afferent* is defined, for the time being at least, as all neural structures in the peripheral and central nervous system which are clearly not efferent (for primitive organisms without an afferent system, the whole organism is presumably involved). *Potentially capable* indicates that R has been reliably elicited in the life history of some representative members of the organism in question; *elicitation* means that R is contingent upon an immediately antedating S.

*(a) Objects and Events.*   Kinesthetic and other interoceptive stimuli, including those produced by thoughts or images, cannot be identified with objects but can be inferred from specific behavioral or physiological events; also the removal of an object—such as food from an established behavior sequence—is an event, rather than an object or a form of physical energy, and this event helps define such removal-produced unconditioned stimuli as aversive stimuli or as frustration-instigation.

*(b) Neural Involvement.*   Although physiologists use the term stimulus to refer to an agent that directly activates a muscle or motor neuron, psychologists have typically reserved the term innervation for this process. This usage is supported by the fact that the laws of behavior for innervation are often not the same as for elicitation by a stimulus, that is, when the afferent nervous system is clearly involved. (For the contrary effects of innervation, see Doty & Giurgea, 1961; Eikelboom & Stewart, 1982; Wagner, Thomas, & Norton, 1967.)

*(c) Elicitation of Responses.*   Defining S in terms of R seems circular, but this is not the case at all. The S in an S-R law is defined independently of *this* R, just as R is defined independently of *this* S. Some R at some time or another, however, has had to be contingent upon this S before S could be classified as such. Today when black versus white is being used as a visual difference (S) for a rat, we are fully aware that black/white discrimination learning was firmly (and independently) established in rats of the same species 60–70 years ago. The R elicited by S can be wholly perceptual or refer to an activity of any sort; but awareness, in any sense, is unnecessary.

*(d) Species (Class of Organisms).*   Frequencies, wave lengths, and intensities that are never responded to are simply not stimuli for the class of organism in question. But what is ultrasound for us is a good S for many other mammals, and this example can be multiplied many times. Ratner emphasized the importance of this definiens with his caveat, "Know your animal" (Denny & Ratner, 1970).

This definition simply reflects the way S is used in the description of experimental procedures in psychological research. The definition implies "once an S, always an S," whether on occasion it elicits an R or not. As a concept with a constant meaning, S can therefore enter into lawful statements with R (in a scientific law, the concepts involved must have a constant meaning). When R is predictable from an S-R law and yet fails to occur in the presence of S, this has no bearing on the definition of S. It simply means that strong, competing stimuli or other relevant variables are present and need to be taken into account when explaining the observed behavior.

## Definition of Response Class (R)

Response class is inferred from three sets of referents: (a) a set of response occurrences for (b) a particular class of organisms in which the identification of the response class is primarily based on (c) some recurrent aspect of the stimulus situation. For example, if a rat (b) is running east (a) toward a place where it has previously found food (c), then R is classified as approach. If the same running behavior (a) is observed in the rat (b) when there is an electrified grid to the west (c), then R is classified as withdrawal (a more specific classification as escape or avoidance would depend upon whether or not the rat had been in contact with the grid).

In the theory, overt response occurrences are necessary as observables to infer a particular response class, but this does not mean that the response class must itself represent overt behavior. Perception, thinking, and relaxation are all legitimate response classes which, in human beings, are typically inferred from verbal report (the response occurrence). For both implicit and explicit behaviors, there is no necessary one-to-one correspondence between response occurrence and response class (a verbal report is not isomorphic with a percept or image). The task of classifying response in animals other than man is somewhat more complex but logically equivalent.

In an everyday example of defining a response class in which, say, an adult cardplayer kicks his adult partner in the shins, the important aspects of the stimulus situation for inferring that the kick in the shins for this class of organism should be classified as illegal signalling behavior are (a) they are playing bridge and (b) it is the kicked person's turn to bid or play a card. In addition, the absence of any facial or verbal expression of hostility by the kicker helps rule out aggression, and the absence of some response occurrence like "Excuse me" from the kicker helps rule out clumsiness. Other information, namely, to the effect that the kicker holds an unusual hand, that the opponents are winning, and that the kicker dislikes to lose is reserved for explaining the kick in the shins. Thus circularity is avoided: The S (at bridge it's the kicked adult's turn to bid) for defining R is different from the S (the kicker holds an unusual hand) in the S-R relation that helps explain R.

A psychologist might be interested in asking the kicker in private why he kicked his partner, and the person might answer, "I thought she might realize I had a very poor hand if I kicked her." Because this verbal report reveals something socially unacceptable about our kicker, the psychologist is very likely to assume that the kicker is telling the truth and thus be more confident about classifying the kick in the shins as illegal signalling behavior than he or she would be with all the other defining characteristics for R that we have just discussed. Thus, inferring a response class from a verbal report may be one of the most accessible and reliable bases for inferring Rs in adult humans. At least, verbal report can hardly be disqualified for use as the principal response occurrence for inferring Rs, whether R is considered cognitive or not.

According to the emphasis of this chapter, the most significant aspect of the recommended definition of R is that emotions, percepts, thoughts, memories, expectancies, and images are no different in their inferential or definitional status than chastising a child or approaching a positive incentive. In all instances, response occurrences are used as referents to help classify the behavior. In fact, for implicit behavior, much of which is typically called cognitive, the psychologist is as dependent on directly observable response occurrences for classifying it as if the behavior being classified were overt. When, for classification purposes, an inference about R is based on a human being's verbal report or on a nonhuman animal's differential response while making a perceptual discrimination, few alternative interpretations of R are usually available, aside from prevarication on the part of the human being. With overt behavior, on the otherhand, a response occurrence such as jumping can be classified innumerable ways, depending on the stimulus situation and the class of organism involved; e.g., jumping as a form of normal locomotion (kangaroo, frog, grasshopper, etc.) or as unusual human locomotion, as in a potato-sack race, jumping to catch the ball or reach an apple, jumping for joy, jumping while dancing, and jumping to demonstrate the point being made here. In other words, one might very well be more confident that a person was seeing or imaging something, from that person's verbal report, than that he was overtly scratching his head in puzzlement (classification of response) when he was overtly hesitant in his speech while attempting to explain a difficult point. Thus, public agreement for classifying so-called cognitive behaviors can often be superior to agreement for classifying overt behaviors, even though the latter are themselves objective.

Admittedly, misclassification of overt behavior is more widespread in the area of personality/social psychology, where good agreement among judges is often the main criterion for response classification, as in classifying some behavior as cooperative, sociable, paranoid, or industrious, than, say, in the area of experimental psychology. But even experimental psychologists have not been immune to misclassifying responses, as for example with the rat's behavior in a T-maze.

In the past, S-R psychologists were inclined to classify a rat's learned response arbitrarily when it was reinforced, say, at the end of the right arm of the T, as a right-turning response. This hasty practice gave rise to the response versus place pseudoproblem. The early work of Tolman and associates (Tolman, Ritchie, & Kalish, 1946, 1947), the analyses of Restle (1957) and of Denny (1967) of response versus place, which included careful observation of all of the rats' response occurrences at the choice point, and experimental data (e.g., Mackintosh, 1965) strongly indicate that what the rat learns first, at any rate, is to approach the place where the incentive is located. Only after many trials, when VTE behavior has completely dropped out and the animal zooms to the right (left) regardless of the orientation of the stem of the maze (i.e., when kinesthetic stimuli from the rat's centrifugal swing in the stem of the T have primary stimulus control), is it legitimate to classify the learned R as a right-turn (left-turn). In other words, all Rs are inferences based only in part on response occurrences, and Rs when appropriately defined are what we are attempting to explain or predict.

## ILLUSTRATIONS OF THE EFFICIENCY OF AN APPROPRIATE S-R SYSTEM

The generic term "memory," if not reified as some sort of mentalistic faculty, refers to the same topical area as retention, but memory presumably includes many response classs that were often ignored or sidestepped when the label of retention was in vogue. Terms such as strategies, mnemonics, working memory, mode of encoding, subjective organization, retrieval, levels of processing, scanning, memories, images, and rotated images are examples of response classes recently in vogue. In short, memory includes a potpourri of implicit behaviors and response-related independent variables that can come into play after the original stimulus (often for a target response) has disappeared, at least momentarily.

### Images as Conditioned Perceptual Responses

The term "memories," on the other hand, refers to rather specific classes of response: verbal representations or sensory images of the original stimulus. It is possible in certain circumstances that the representation is purely a neural representation of the stimulus (NRS), but only when the organism is completely unaware of the cue and cannot be made aware of it. Then one wonders if such effects would qualify as memories.

For most of the work reported in this volume, the memories qualify as images. The tasks presented to the pigeon, monkey, or rat are typically visual discrimination problems of one sort or another in which the animal must clearly attend to a particular element, or set of elements, in the stimulus situation. There is nothing subliminal or subtle about the relevant cue. Thus if the critical cue is clearly a percept, it stands to reason that the "memory," to function at all, must be as similar as possible to the percept, that is, a representative visual image.

There are other reasons, as well, for positing images as the chief sort of memories in animals other than humans. On general systematic or theoretical grounds, it makes supreme sense. There is absolutely no theoretical reason why animals devoid of language should be without imagery. In the behavioristic mode, images can be conceived of as first- and second-order conditioned perceptual responses to a set of conditioned stimuli, quite liable to both mediated and primary generalization. Thus, since such animals clearly perceive, they are eminently capable of having images. Images fit well into the category of conditioned responses because, like all conditioned responses, they are not identical to the unconditioned response (in this case the percept). Images are typically vague, incomplete, and less vivid when compared to percepts. Leuba (1940) demonstrated such conditioning in college-student subjects for cutaneous, olfactory, auditory, and visual images using a neutral stimulus like a bell as the conditioned stimulus. The subjects, who were typically hypnotized during conditioning and tested when awake, were dumbfounded when the images occurred.

Currently, Sheikh (1983) is a vigorous proponent of the view that images are conditioned sensations (conditioned perceptual responses), and Finke (1980), among others, has demonstrated in several intricate ways the degree to which imagery and perception are equivalent in human beings. At both the human and nonhuman level, imagery can constitute a sizable portion of what is ordinarily meant by thinking. So one needs to emphasize that the cognitive "revolution," independent of its conceptual trappings, is a salutary, overdue movement in terms of the variables manipulated, analyses made, and response classes investigated.

## Forgetting and Extinction

In the alternate S-R framework, there are two main sources of forgetting: interference (both PI and RI) and stimulus change or generalization decrement. The latter can also produce incidental interference if stimulus change involves the introduction of new competing stimuli. Stimulus change, since it weakens the original response tendency through generalization decrement, can generally enhance PI and RI: An interfering response can compete more effectively, causing blocking or forgetting, when the to-be-remembered response is weaker. The role of interference is nicely illustrated in animal memory studies by Kendrick, Rilling, & Stonebraker (1981), in which forgetting (poor delayed matching to sample performance) is shown to be due to responses that are incompatible with accurate key-pecking performance.

This analysis holds for short-term as well as for long-term memory. For short-term situations, the fading of original stimuli, often referred to as a memory trace, can represent both the introduction of new stimuli (very likely producing both incidental interference and generalization decrement) and the loss of original stimuli (generalization decrement effect). The variables that produce forgetting are much the same as those that are assumed to be mainly involved in experimental extinction. In general, competing responses, which are frequently elicited by the frustration of nonreward, bring about the extinction of a learned response (Adelman & Maatsch, 1955; Denny, 1971B; Hilgard & Marquis, 1935; Johnson & Denny, 1982; Wong, 1971). Relief or relaxation seem to be the interfering responses that contribute to the extinction of escape-avoidance behavior or fear (Delprato & Dreilinger, 1974; Denny, 1971a; Grelle & James, 1981). Moreover, generalization decrement, in extinction as in forgetting, clearly enhances the decremental effects (e.g., Denny, 1971b; Hurwitz & Cutts, 1957; Reynolds 1945; Sheffield, 1950; Stanley, 1952; Teichner, 1952; Welker & McAuley, 1978).

Without interference or stimulus change, forgetting of well-established habits from the mere passage of time should be minimal. That this is the case for a variety of organisms has been shown in a variety of learning situations (Brogden, 1951; Gagné, 1941; Wendt, 1937). And a recent observation describes what is probably the longest retention period in a nonmammalian animal. After 12 years, a pigeon showed retention of stimulus control for keypecking as well as for schedule effects of the FR and FI components of the multiple schedule used in original training (Donahoe & Marrs, 1982).

## The Role of Contextual Cues

The role of retrieval cues, reinstatement, and contextual cues is well accepted in the analysis of memory effects and is a clear part of the jargon of the cognitive movement. But these notions also fit an S-R framework. In 1945–1946, I was inspired by one of Spence's seminar suggestions that the reinstatement of some of the cues present during acquisition when instituted just prior to recall should facilitate recall. This hypothesis was tested in a paired-associate nonsense syllable learning experiment with college students at the University of Oklahoma. All subjects learned the 10-item list to a criterion of one errorless trial and were tested for retention with several recall trials 48 hours later. There were three experimental, or reinstatement, groups and one control (no reinstatment). One reinstatement group just prior to recall was given the list of stimulus words to mull over for 2 minutes, another just the list of response words to mull over, and the third reinstatement group both the S and R words but explicitly unpaired. The S group and R group both recalled more items than the control, but only the S-R group was significantly superior to the control ($p = .02$). In addition, the control group improved much more on the second and third trials than the experimental groups, presumably receiving much of the reinstatment of cues on the first recall trial. (The improvement from first to second recall trial was considerably greater than any one-trial increment by any group during original acquisition.)

The argument in this study was that subjects during acquisition trials were seeing the S and R words on every trial and saying aloud all of the R words by the final trial, and all of these elements constituted a critial portion of the stimulus context in which learning of the specific pairs took place. Ordinarily on the first recall trial, a subject is handicapped by not having the full stimulus context available that was present during original learning, and thus some of the forgetting that typically occurs is presumably due to generalization decrement. Mulling over the nonsense syllables reinstates contextual cues present during learning. Thus generalization decrement is reduced, and the handicap is minimized.

When the written report for this experiment was submitted for publication, I realized reviewers would probably interpret the reinstatement procedure as an opportunity to rehearse the original material, so a note was included to the effect that subjects could only correctly rehearse pairs that they specifically recalled during the short mulling-over period, which of course they would recall anyway on the first recall trial (and besides there was no feedback). Nothing in the mulling-over period specified anything about correct or incorrect pairs. The more reasonable interpretation, but not to the journal reviewers, seemed to be in terms of cue reinstatement. The main reason for describing this unpublished retention experiment, regardless of whether it was sound or not, is to emphasize that it evolved from a straight S-R framework nearly 40 years ago, and before the availability of the information-processing literature. Another reason is to provide a lead for pointing out how important these same ideas currently are in the area of traditional S-R research, e.g., with rats in a straight runway.

Jobe, Mellgren, Feinberg, Littlejohn, and Rigby (1977) demonstrated that retrieval cues are essential for simple alternation patterning to occur in a rat with one trial a day in a runway (inhibiting on nonreinforced days and running fast on alternate reinforced days). A large incentive in the goal box on reinforced days was also found to be a critical factor, presumably to make both reinforced and nonreinforced events highly salient. For patterning to occur, the start box had to be similar to the goal box so as to trigger the "memory" of what had occurred in the goal box on the previous day (presumably some sort of olfactory-gustatory-visual image following reinforced trials and some sort of emotional or frustration-instigated response following nonreinforced trials). And, using reinstatement techniques similar to Jobe et al., Jobe, Mays, and Mellgren (1982) demonstrated that retrieval cues are needed for a good partial reinforcement extinction effect (PRE) even with an ITI of only 15 s. That is, memories, as is true of all other response classes, are cue-dependent even at very short intervals.

## Long-delay Learning

As soon as we assume that animals other than humans can have images, it becomes possible to explain the occurrence of learning with long conditioned stimulus-unconditioned stimulus (CS-US) intervals, as in taste-aversion conditioning (e.g. Barker, Best, & Domjam, 1977; Garcia & Koelling, 1966) and very possibly for long delays of reinforcement (Lett, 1973). Taste-aversion conditioning has the following special characteristics (Riley & Baril, 1976), which are all entirely consistent with an interpretation of long-delay conditioning in terms of imagery:

1. Very long CS-US intervals, even as long as a day, yield conditioning.
2. Considerable learning takes place in one trial.
3. Especially when the CS-US interval is fairly long, the Garcia effect holds for species like the rat: A taste or olfactory stimulus is a good CS when the aversive US is a toxic substance that causes illness but a poor CS when the aversive US is electric shock, and, vice versa, a visual or auditory stimulus is a poor CS for a toxic US but a good CS when the US is shock. For the visually oriented bird, however, a visual stimulus especially when the food itself is colored, is an excellent CS for taste aversion conditioning.
4. So-called backward conditioning can occur readily.
5. The taste CS must be relatively novel for good conditioning to occur.
6. The introduction of other novel tastes during the long-delay interval disrupts conditioning to the target CS flavor (Kalat & Rozin, 1971; Revusky, 1971).
7. Reconditioning an aversion to a taste CS after extinctin of this aversion is poor or nonexistent (Danguir & Nicolaidis, 1977).

Imagery could operate in the taste-aversion conditioning situation in the following manner. After ingesting and tasting the CS substance for a number of minutes, during the original conditioning trial, the subject typically eats or drinks nothing until made ill by an injection of the toxic US. When illness occurs, the malaise or gastronomical upset presumably elicits images of the most recent gastronomical events (probably salient events for most animals), which would *include* gustatory and probably olfactory images of the taste CS, but not visual or auditory images for most animals (point 3 above). Thus the gustatory or olfactory image CS would actually be contiguous with the UR, or nausea, and good conditioning would be expected (the long delay of point 1 above is not really long at all and the relevance of points 3, 5, and 6 is obvious).

Also the image could be elicited a number of times while the subject is ill, actually resulting in more than a single conditioning trial for one pairing of CS and US (point 2 above). This concept also explains the so-called backward conditioning of point 4 above: Long-lasting nausea (Unconditional Response [UR] would take a while to occur and would certainly follow the taste CS even though the toxic substance (US) preceded it during the conditioning trial, and the image CS would readily accompany or immediately precede nausea (that is, forward conditioning rather than backward conditioning actually prevails because the forward contiguous association in question is S with R, i.e., image CS with UR or nausea).

A visual or auditory CS for a rat, through neural connections or past experience, is not intimately associated with gastronomical events; thus visual or auditory images would *not* be elicited by nausea and therefore could not mediate aversive conditioning with a long CS-US interval (point 3 above). However, a visual-auditory CS *can* mediate aversive conditioning for a US that produces nausea if the CS-US interval is conventionally short, that is, when images are not involved. For a bird that constantly searches for food visually, it would seem reasonable for it to have an image of any novel visual stimulus that preceded nausea—even by several hours—and thus show good aversive conditioning to such a visual CS (latter half of point 3).

To be most effective as a CS, the gustatory image must be uniquely elicited by the nausea, and for this to occur the CS must be novel and unaccompanied by other novel stimuli. Such a CS would *not* be conditioned to competing responses and would be distinctive enough to be remembered hours later as an image when the illness occurred (points 5 and 6 above). Finally, if the true CS is an image and the aversive reaction to the generalized taste CS were extinguished, then the association between nausea and gustatory image would also undergo counterconditioning (or extinction). Thus the gustatory image would not be elicited by nausea during reconditioning trials; and reconditioning, with the actual CS absent (the gustatory image extinguished), would be very poor or nonexistent (point 7 above). In ordinary reconditioning situations, the CS is definitely present and thus reconditioning is typically rapid.

## The Differential Outcome Effect

Finally, let us discuss the differential outcome effect (DOE) in the learning of two-choice conditional discriminations (Brodigan & Peterson, 1976; Carlson & Wielkiewicz, 1972; Peterson, Wheeler, & Armstrong, 1978; and Trapold, 1970). In the typical paradigm, a pigeon or rat is presented with one of two conditional cues, $S_1$ or $S_2$, to which it makes one of two responses, $R_1$ or $R_L2$. When the outcomes for the correct response to $S_1$ and $S_2$ differ from each other, learning is clearly facilitated as compared to nondifferential outcomes (the traditional procedure). The differential outcomes studied include food versus water, both food but qualitatively different, same food but differing in amount, and even one outcome reinforcement and the other nonreinforcement or a neutral stimulus like a tone (Peterson, Wheeler, & Trapold, 1980).

The main interpretation of facilitative DOE involves invoking differential expectancies for the different outcomes. In the S-R mode, $E_1$ and $E_2$, the two different expectancies (Rs with their attendant and characteristic stimulus properties), are conditioned to $S_1$ and $S_2$, respectively, and as $E_1$ and $E_2$, they antedate $R_1$ and $R_2$. The unique interoceptive properties of $E_1$ and $E_2$ thus augment the exteroceptive differences between $S_1$ and $S_2$ and more precisely elicit the correct response ($R_1$ or $R_2$). To the extent that there is a delay between $S_1$ or $S_2$ and the availability of $R_1$ and $R_2$, $E_1$ and $E_2$ as intervening and longer-lasting stimulus events can mediate correct responding and enhance conditional discrimination learning even more than without such a delay (Peterson & Trapold, 1980).

Such expectancies can probably be conceived of as images, emotional responses, or as $r_g-s_g$ and $r_f-s_f$ in the Hull-Spence-Amsel tradition; but what is most important is that the expectancies, as shown by the DOE data, are quite specific to their outcomes. Thus the response class designation of expectancy as it applies to human beings seems equally appropriate for many other animals; and a flexible S-R framework is quite comfortable with this state of affairs. (See also Chapter 1 for a more restricted view of the S-R position).

## CONCLUSION

When stimulus (S) and response class (R) are appropriately defined, admitting every sort of afferent and efferent event to the theory of behavior, there is no need to discard S-R concepts and theorizing in favor of a cognitive position. Such an S-R framework has several advantages. It represents a consistent system that relates to all behavior in an integrative fashion, it prevents the growth of mentalistic ideas, concepts, and entelechies by providing an objective basis for analyzing behavior, and it ensures the retention of everything useful that we discovered in the past from an S-R perspective. Moreover, all psychologists, including cognitivists, use input and output concepts extensively, and usually some form of stimulus and response. So what do we gain by dismissing S-R psychology? In fact, most of the authors of the chapters in this volume were either S-R psychologists just a few years ago or still are. Those who openly discard S-R probably view it from a narrow, traditional perspective that does not permit some important problems to be researched.

From the present perspective, an important difference between the traditional S-R view and the cognitive position is simply the kinds of problems currently being researched. Attention to a new set of problems has brought about the introduction of new concepts, which, according to this chapter, mainly refer to response classes that have either been ignored for decades or are only now being considered. That is, this difference evaporates when S and R are appropriately defined.

A final note: Averill (1983) suggests for the study of human behavior that the cognitive wave is on the wane. Again, what this seems to mean is that a number of psychologists are turning their interests away from problems of information processing toward problems of emotional behavior in human interactions.

# REFERENCES

Adelman, H. M., & Maatsch, J. L. Resistance to extinction as a function of the type of response elicited by frustration. *Journal of Experimental Psychology,* 1955, *50,* 61–65.

Averill, J. R. Studies on anger and aggression. *American Psychologist,* 1983, *38,* 1145–1160.

Barker, L. M., Best, M. R., & Domjan, M. *Learning mechanisms in food selection.* Waco, TX: Baylor University Press, 1977.

Brodigan, D. L., & Peterson, G. B. Two-choice conditional discrimination performance of pigeons as a function of reward expectancies, prechoice delay, and domesticity. *Animal Learning and Behavior,* 1976, *4,* 121–124.

Brogden, W. J. Animal Studies of learning. In S. S. Stevens (Ed.), *Handbook of experimental psychology.* New York: Wiley, 1951.

Carlson, J. G., & Wielkiewicz, R. M. Delay of reinforcement in instrumental discrimination learning of rats. *Journal of Comparative and Physiological Psychology,* 1972, *81,* 365–370.

Danguir, J., & Nicolaidis, S. Lack of reacquisition in learned taste aversion. *Animal Learning and Behavior,* 1972, *5,* 395–397.

Delprato, D. J., & Dreilinger, C. Backchaining of relaxation in the extinction of avoidance. *Behavioral Research and Therapy,* 1974, *12,* 191–197.

Denny, M. R. A theoretical analysis and its application to training the mentally retarded. In N. R. Ellis (Ed.), *International review of research in mental retardation: Vol 2.* New York: Academic Press, 1966.

Denny, M. R. A learning model. In W. C. Corning & S. C. Ratner (Eds.), *the Chemistry of Learning.* Lincoln, NE: University of Nebraska Press, 1967.

Denny, M. R. Relaxation theory and experiments. In F. R. Brush (Ed.), *Aversive conditioning and learning.* New York: Academic Press, 1971a.

Denny, M. R. A theory of experimental extinction and its relation to a general theory. In H. A. Kendler & J. T. Spence (Eds.), *Essays in neobehaviorism.* New York: Appleton-Century-Crofts, 1971b.

Denny, M. R., & Ratner, S. C. *Comparative psychology: Research in animal behavior.* Homewood, IL: Dorsey, 1970.

Donahoe, J. W., & Marrs, D. P. 12-year retention of stimulus and schedule control. *Bulletin of Psychonomic Society,* 1982, *19,*(3), 184–186.

Doty, R. W., & Giurgea, C. Conditoned reflexes established by coupling electrical excitation of two cortical areas. In J. F. Delafresaye (Ed.), *Brain mechanisms and learning.* Oxford, UK: Blackwell, 1961.

Eikelboom, R., & Stewart, J. Conditioning of drug-induced physiological responses. *Psychological Review,* 1982, *89,* 507–528.

Finke, R. A. Levels of equivalence in imagery and perception. *Psychological Review,* 1980, *87,* 113–132.

Gagné, R. M. The retention of a conditioned operant response. *Journal of Experimental Psychology,* 1941, *29,* 293–305.

Garcia, J., & Koelling, R. A. Relation of cue to consequence in avoidance learning. *Psychonomic Science,* 1960, *4,* 123–124.

Grelle, M. J., & James, J. H. Conditioned inhibition of fear: Evidence for a competing response mechanism. *Learning and Motivation,* 1981, *12,* 300–320.

Guilford, J. P. Cognitive psychology's ambiguities: Some suggested remedies. *Psychological Review,* 1982, *89,* 48–59.

Hilgard, E. R., & Marquis, D. G. Acquisition, extinction, and retention of conditioned lid responses to light in dogs. *Journal of Comparative Psychology,* 1935, *19,* 29–58.

Hurwitz, H. M. B., & Cutts, J. Discrimination and operant extinction. *British Journal of Psychology,* 1957, *48,* 90–92.

Jobe, J. B., Mays, M. Z., & Mellgren, R. L. Reinstatement of retrieval cues, intertrial interval, and resistance to extinction. *Bulletin of Psychonomic Society*, 1982, *19*(3), 163–164.

Jobe, J. B., Mellgren, R. L., Feinberg, R. A., Littlejohn, R. L., & Rigby, R. L. Patterning, partial reinforcement, and N-Length effects at spaced trials as a function of reinstatement of retrieval cues. *Learning and Motivation*, 1977, *8*, 77–97.

Johnson, C. M., & Denny, M. R. Inhibition of performance as a function of withdrawal from nonreinforcement in a partial reinforcement situation. *Psychological Record*, 1982, *32*, 315–327.

Kalat, J. W., & Rozin, P. Role of interference in taste-aversion learning. *Journal of Comparative & Physiological Psychology*, 1971, *77*, 192–197.

Kendrick, D. F., Rilling, M., & Stonebraker, T. B. Stimulus control of delayed matching in pigeons: Directed forgetting. *Journal of the Experimental Analysis of Behavior*, 1981, *36*, 241–251.

Lett, B. T. Delayed reward learning: Disproof of the traditional theory. *Learning and Motivation*, 1973, *4*, 237–246.

Leuba, C. Images as conditioned sensations. *Journal of Experimental Psychology*, 1940, *26*, 345–351.

Lewis, D. J. *Scientific principles of psychology.* Englewood Cliffs, NJ: Prentice-Hall, 1963.

Mackintosh, N. J. Overtraining, transfer to proprioceptive control, and position reversal. *Quarterly Journal of Experimental Psychology*, 1965, *17*, 26–36.

Pear, J. J. Relative reinforcements for cognitive and behavioral terminologies. In Symposium Proceedings. On cognitive and behavioral orientation to the language of behavior analysis: Why be concerned over the differences? *Psychological Record*, 1983, *33*, 3–30.

Peterson, G. B., & Trapold, M. A. Effects of altering outcome expectancies on pigeon delayed conditional discrimination performance. *Learning and Motivation*, 1980, *11*, 267–288.

Peterson, G. B., Wheeler, R. L., & Armstrong, G. D. Expectancies as mediating the different-reward conditional discrimination performance of pigeons. *Animal Learning and Behavior*, 1978, *6*, 276–285.

Peterson, G. B., Wheeler, R. L., & Trapold, M. A. Enhancement of pigeon's conditional discrimination performance by expectancy of reward. *Animal Learning and Behavior*, 1980, *8*, 22–30.

Restle, F. Discrimination of cues in mazes: A resolution of the "Place-vs.-response" question. *Psychological Review*, 1957, *64*, 217–228.

Revusky, S. H. The role of interferance in association over a delay. In W. K. Honig & P. H. R. James (Eds.), *Animal memory.* New York: Academic Press, 1971.

Reynolds, B. Extinction of trace conditioned responses as a function of the spacing of trials during the acquisition and extinction series. *Journal of Experimental Psychology*, 1945, *35*, 81–95.

Riley, A. L., & Baril, L. L. Conditioned taste aversions: A bibliography. *Animal Learning and Behavior*, 1976, *4*(1b), 15–135.

Schnaitter, R. A new fable for Aesop. In symposium Proceedings. On cognitive and behavioral orientations to the language of behavior analysis: Why be concerned over the differences? *Psychological Record*, 1983, *33*, 3–30.

Sheffield, V. F. Resistance to extinction as a function of the distribution of extinction trials. *Journal of Experimental Psychology*, 1950, *40*, 305–313.

Sheikh, A. A. *Imagery, current theory, research and applications.* New York: Wiley, 1983.

Stanley, W. C. Extinction as a function of the spacing of extinction trials. *Journal of Experimental Psychology*, 1952, *43*, 249–260.

Teichner, W. H. Experimental extinction as a function of the intertrial intervals during conditioning and extinction. *Journal of Experimental Psychology*, 1952, *44*, 170–178.

Tolman, E. C., Ritchie, B. F., & Kalish, D. Studies in spatial learning: II Place learning versus response learning. *Journal of Experimental Psychology*, 1946, *35*, 221–229.

Tolman, E. C., Ritchie, B. F., & Kalish, D. Studies in spatial learning: V Response learning versus place learning by the non-correction method. *Journal of Experimental Psychology*, 1947, *37*, 285–292.

Trapold, M. A. Are expectancies based on different positive reinforcement events discriminably different? *Learning and Motivation*, 1970, *1*, 129–140.

Wagner, A. R., Thomas, E., & Norton, T. Conditioning with electrical stimulation of motor cortex: Evidence of a possible source of motivation. *Journal of Comparative & Physiological Psychology* 1967, *64*, 191–199.

Wasserman, E. A. Is cognitive psychology behavioral? In Symposium Proceedings. On cognitive and behavioral orientations to the language of behavior analysis: Why be concerned over the differences? *Psychological Record*, 1983, *33*, 3–30.

Welker, R. L., & McAuley, K. Reductions in resistance to extinction and spontaneous recovery as a function of changes in transportational and contextual stimuli. *Animal Learning and Behavior*, 1978, *6*, 451–457.

Wendt, G. R. Two and one-half year retention of a conditioned response. *Journal of General Psychology*, 1937, *17*, 178–181.

Wong, P. T. T. Coerced approach to shock punishment of competing responses and resistance to extinction in the rat. *Journal of Comparative Physiological Psychology*, 1971, *76*, 275–281.

# II    MEMORY PROCESSES

# Prospection and Retrospection as Processes of Animal Short-Term Memory

## 4

Edward A. Wasserman
*The University of Iowa*

Writing in 1967, Jerzy Konorski observed that most of the problems and procedures of the behavioral sciences "concern the formation and properties of stable memory traces involved in conditioning and learning, whereas the methods devoted to the problems of recent memory have been relatively scarce [p. 490]." Whatever the reasons for this imbalance in emphasis, Konorski felt that it was "in clear disproportion to the important role which is played by the phenomena of recent memory both in our own mental processes and in those of other animals [p. 490]."

Were Konorski still alive, he would have been pleased to see the current surge of interest in animal short-term memory, stimulated in part by renewed attention to the comparative analysis of cognition (Wasserman, 1981, 1982, 1983). We are still some distance away from incorporating the many facts of animal short-term memory into a coherent theoretical system, such as the synthesis that Konorski (1967) attempted for phenomena of long-term memory. Nonetheless, it is not too early to propose at least provisional accounts of how animal behavior may be controlled by recent, but absent events. Although probably incorrect or incomplete, such tentative interpretations may give us new ways to think about old issues; they may also suggest novel problems for experimental investigation. Konorski offered an original analysis of short-term memory, which may well be worthy of much greater attention than it has received to date. This chapter presents Konorski's proposal, evaluates its merits in the light of existing evidence, and looks ahead to further empirical tests.

At the heart of Konorski's analysis of short-term memory phenomena is the distinction between prospection and retrospection as processes mediating delayed control by prior events. To explicate this distinction (Konorski, 1967), consider an example: I attend a scientific meeting with several lectures to hear and people to meet. The planned series of times, people, and places serves as the guide to my convention activities. "All these and a host of other future tasks are held in my transient memory store, and I direct my behavior in such a way as to fulfill them. After this is done, their memory is no longer preserved and they 'make room' for other images [p. 494]." Because the plan of action is for one time only, Konorski reasoned that its memory is of the short-term variety. And because memory here involves plans for future actions, he suggested that this showed memory in its *prospective* aspect.

To continue the example, it would be disadvantageous if I were to return to particular places intending to see people already seen and hear lectures already heard. To avoid inefficient repetition, I must also remember finished activities; this, Konorski (1967) proposed, showed memory in its *retrospective* aspect. "Thus our transient memory plays no lesser role in remembering the already completed actions than in remembering those actions which have to be fulfilled in the future. Whereas the impairment of transient memory involved in the latter actions leads to the failure in planning future behavior, the impairment of the memory of fulfilled actions leads to their perseveration [p. 495]."

In outlining the distinction between prospection and retrospection, it is apparent that Konorski (1967) at least tentatively believed that they were separate psychological processes, entailing different neurophysiological mechanisms. How successfully these two hypothesized processes can be separated is in large part the task now before us. A potentially useful starting point for this consideration involves careful examination of some basic paradigms that have been devised for studying short-term memory in animals.

## SHORT-TERM MEMORY PARADIGMS

Although not designed to distinguish the involvement of different memory processes, some short-term memory paradigms would seem to encourage prospective or retrospective mediation for their solution. Discussion of these paradigms may provide a better sense of the prospection/retrospection dichotomy within the context of animal behavior.

### Prospection

Konorski (1967) identified Hunter's (1913) delayed-response paradigm as one especially prone to prospective mediation. In the delayed-response paradigm, one of several possible food sites is baited in the view of the subject. After a delay, the subject is permitted to select one of the alternative response locations. Correct choice in this task requires a form of memory, as the animal's response selection must be based upon site baiting some time earlier. But is this memory the result of residual stimulation occasioned by baiting as implied by retrospective mediation or is it the result of anticipatory action plans as implied by prospective mediation? Konorski believed the latter; "the corresponding cue . . . provided by the . . . image of the itinerary to a given feeder [p. 499, italics removed]."

It is a frequent finding in research using the delayed-response paradigm (Fletcher, 1965; Hunter, 1913) that animals may adopt a persistent orientation to the baited food site during the delay interval. Although such orientation need not be necessary for accurate test performance, it is at least consistent with Konorski's (1967) proposal that an anticipated plan of action is mediating correct performance rather than some durable sensory trace of earlier site baiting. Looking ahead to where to respond should keep the animal oriented toward the baited site, whereas residual aftereffects of viewing the baiting would not seem to have clear implications for delay-interval orientation.

The delayed-response paradigm is not the only one that may be conducive to prospective mediation. In this class, Konorski (1959) also placed Pavlov's (1927) conditioned-inhibition paradigm. Here, the conditioned stimulus (CS) is followed by the unconditioned stimulus (US), but only on those occasions when the CS is presented alone; if it is preceded by the conditioned inhibitor (CI), the CS is not followed by the US. Konorski stated: "It would seem that the recent memory of the conditioned inhibitor is indispensable for the solution of this task. But here again, according to our experience, the conditioned inhibitor evokes a quite ostensible negative bodily attitude which lasts for some time and is the very cause of the negative response to the conditioned stimulus acting against its background [p. 116]." Although the negative bodily attitude of which Konorski spoke is not a plan of action, it is better conceived of as a reaction to anticipated nonreinforcement than as a result of perseverative sensory stimulation. Thus, prospection is a more likely mediating process of CI than is retrospection.

Finally, Konorski (1967) considered what he termed simple delayed-reactions (not to be confused with the previously discussed delayed-response paradigm; Konorski & Lawicka, 1959) to be prospectively mediated. In this paradigm, the CS is followed by the US only when a conditioned excitor (CE) precedes the CS; if the CS alone is presented, no US is delivered. Thus, the simple delayed-reaction paradigm is the converse of the CI paradigm (also see Bottjer & Hearst, 1979; Terry & Wagner, 1975); as such, a positive bodily attitude or set may here be evoked by the reinforcement-paired CE, which leads to enhanced responding to the CS relative to CS-alone presentations.

## Retrospection

In an attempt to discourage prospective mediation of delayed-stimulus control, Konorski (1959) devised a new method for investigating short-term memory in animals. The basic plan of this new paradigm was to present pairs of stimuli to animals and to follow pairs with reinforcement only if they involved the same stimulus; if, however, pairs involved different stimuli, reinforcement was withheld. Were animals to respond differently to the second stimulus of a pair depending upon the identity of the first, Konorski reasoned that such differential performance would have to involve retrospective mediation: "Only a comparison of the second stimulus with the first can provide the animal with the clue indicating how to react. The correct response is possible only if the animal retains the trace of the first stimulus when the second is acting [p. 117]."

As we will see later, Konorski (1959, 1967) may have been a bit hasty in rejecting prospection as a mediator of delayed-stimulus control in his new short-term memory procedure (and in related procedures as well). Nevertheless, Konorski's (1959) successive matching-to-sample paradigm has emerged as an elegant and versatile method of investigating short-term memory in animals (see Nelson & Wasserman, 1978; Wasserman, 1976; Wasserman, Grosch, & Nevin, 1982).

Another variant of matching-to-sample may also be preferentially mediated by retrospection. This is the familiar delayed choice matching-to-sample paradigm (Blough, 1959). In this paradigm, the correct choice from two or more alternative test stimuli depends upon the prior sample stimulus. Only if the subject picks the alternative that matches the sample is reinforcement given; incorrect selections fail to produce reinforcement. Again, because a comparison of the test alternatives with some residue of the sample stimulus seems likely, Konorski would propose that retrospection is more likely to mediate correct performance than is prospection.

The same logic applies as well to other variants of the delayed matching-to-sample paradigms. Particularly important to later discussion is so-called symbolic matching-to-sample (Cumming & Berryman, 1965). Here, the samples are selected from a set of stimuli which are physically different from the set of possible test stimuli. Thus, for example, the sample stimuli may involve patches of different colors, whereas the test stimuli may entail different geometric forms. With a set size of two, four color-form sequences are possible in Konorski's (1959) successive matching-to-sample paradigm: two of which (Color 1-Form 1 and Color 2-Form 2) can be associated with reinforcement and two of which (Color 1-Form 2 and Color 2-Form 1) can be associated with nonreinforcement (also see DeLong & Wasserman, 1981; Honig & Wasserman, 1981; Nelson & Wasserman, 1981).

## IS PROSPECTION MORE DURABLE THAN RETROSPECTION?

If prospection and retrospection are actually distinguishable psychological processes, then perhaps some independent variables will have different behavioral effects on tasks that are selectively or preferentially mediated by these processes. Just such a possibility is suggested by an experiment of Smith (1967).

In a provocative, but rarely cited study, Smith (1967) compared the effects of the sample-test retention interval on discriminative performance in two short-term memory tasks. In one task, the orientation of a triangle on the center key signalled which of two side keys the pigeons should later peck to procure reinforcement; one orientation cued a later left key peck and the other cued a later right key peck. This delayed *simple* discrimination (Honig & Wasserman, 1981) is one that Konorski might well have judged would be solved on the basis of prospection, the sample here enabling the birds to anticipate the position of the key to be pecked on the upcoming choice test. In the second task, the pigeons were to choose the side key that displayed the same line orientation stimulus that had earlier been presented on the center key as the sample stimulus. Because the two line orientations were equally presented as the sample stimulus, and because the choice tests involved randomized presentation of the two line tilt stimuli on the side keys, this delayed *conditional* discrimination (Honig & Wasserman, 1981) precluded the birds from anticipating the position of the key to be picked on the choice test. Thus, this task is one which Konorski should have proposed would be solved on the basis of retrospection, the sample here having to be retained in its sensory aspect until the choice test.

The results showed quite clearly that, as the retention interval was increased from 0 s to 5 s, discriminative performance fell more precipitously for the conditional discrimination than for the simple discrimination. Perhaps, then, prospection is a more durable mediating process than retrospection, and perhaps the simple discrimination involves only the former process and the conditional discrimination involves only the latter process.

In an effort to extend the generality of Smith's (1967) findings, Honig and Wasserman (1981) also studied the effects of the sample-test retention interval on delayed discriminations of the simple and conditional varieties. Here, however, discriminative performance was assessed by key pecking rates on go/no go tests that involved the presentation of only one stimulus at a time. Thus, in the simple task, the particular sample color signalled whether reinforcement or nonreinforcement would follow the line orientation test stimulus; the specific line orientation that was presented was irrelevant to the trial outcome. In the conditional task, both the sample and the test stimuli conjointly provided information about reinforcement and nonreinforcement.

Once again, discriminative test performance dropped faster for the conditional task than for the simple task as the retention interval was increased from 0 s to 25 s (see Figure 4.1).[1] We may then surmise that anticipated reinforcement or nonreinforcement in the simple task can provide more durable mediational information than can residual afferent impulses occasioned by different key colors in the conditional task.

Pursuing the logic that some independent variables may have different effects on tasks that are differentially mediated by prospection and retrospection, researchers (D. Chatlosh, V. Guttenberger, and S. Morgan) in my laboratory recently completed an experiment concerned with the effects of sample duration on delayed simple and conditional discriminations in pigeons.

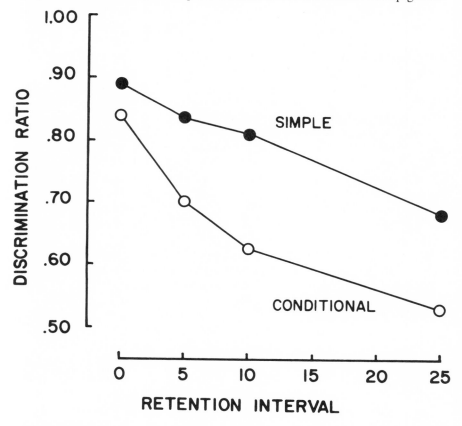

FIG. 4.1.    Mean discrimination ratios of eight pigeons simultaneously trained on delayed simple and conditional go/no go discriminations.

[1]The data for the figure come from Experiment 2 of Honig and Wasserman (1981), and are the average of Blocks 15 (0-, 5-, and 10-s retention intervals) and 16 (5-, 10-, and 25-s retention intervals) of training (blocks of training were 4 days long).

This study was a systematic replication of Smith's (1967) choice-matching study, but it used the same sets of sample and test stimuli as Experiment 2 of Honig and Wasserman (1981). Figure 4.2 documents the more pronounced forgetting function for the conditional discrimination than for the simple discrimination, depicted earlier in Figure 4.1. It further shows that increases in sample duration had a much larger effect on delayed discriminative performance in the conditional task than in the simple task.[2] If it is the case that retrospection involves perseverative sensory processes (Roberts & Grant,

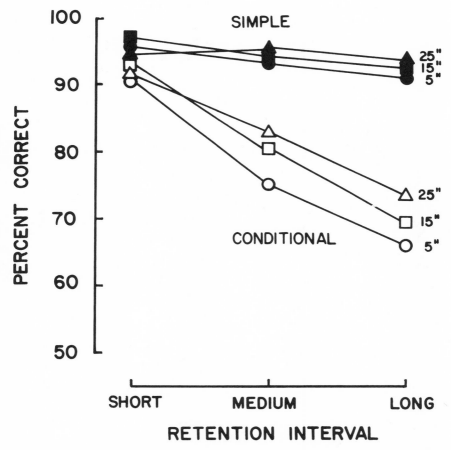

FIG. 4.2.   Mean choice accuracy of four pigeons simultaneously trained on delayed simple and conditional discriminations.

[2]The data for the figure come from unpublished work by Chatlosh, Guttenberger, and Morgan, and are the average of 12 days at each of three sample durations (5, 15, and 25 sec). The "short" retention interval was always 0 sec; "medium" and "long" retention intervals were adjusted for individual birds, and ranged from .5 sec to 10 sec.

1976), then increases in the duration of the sample stimulus should have a facilitative effect on memory performance; on the other hand, if prospection requires only enough stimulus exposure to permit the animal to make an anticipatory test response, then little effect of sample duration should be seen. Our results are consistent with these proposals. Unfortunately, they are also consistent with the less exciting possibility that there was a "ceiling" on performance under the simple discrimination task. We are currently replicating this experiment, while equating the performance of birds under the simple and conditional problems, in hopes of eliminating any ceiling effect on simple discrimination performance.

Before proceeding to additional evidence on the relative persistence of prospection and retrospection, a particular methodological feature of the Honig and Wasserman (1981) study (and our later follow-up) deserves special note here. Whereas Smith (1967) had confounded the specific stimuli he employed as samples in the two discrimination problems—triangles were always samples in the simple task and lines were always samples in the conditional task—Honig and Wasserman made sure that the same physical stimuli served equally often as samples in the two tasks. Such control eliminates a trivial reason for Smith's results: namely, that the differently oriented triangles he used were more easily discriminated and remembered than the differently angled lines he employed. Quite apart from methodological matters, however, the fact that Honig and Wasserman repeated Smith's results with equally discriminable and rememberable sample stimuli suggests that a rather important psychological question must be answered: How can the same physical stimuli occasion memories that fade at different rates? While silent on the question of whether only one or both the simple discrimination and the conditional discrimination entail prospective mediation, the differently sloped memory functions from these two tasks (Figures 4.1 and 4.2) require that we reject the hypothesis that both delayed discriminations are mediated retrospectively.

Further evidence on the relative longevity of prospection and retrospection comes from a line of work begun by Trapold (1970) and importantly extended by Peterson and his associates (Brodigan & Peterson, 1976; Peterson & Trapold, 1980; Peterson, Wheeler, & Armstrong, 1978; Peterson, Wheeler, & Trapold, 1980). Much of that research examined delayed symbolic matching-to-sample performance under two different conditions. In one, distinctive trial outcomes (e.g., food and water) were differentially associated with particular sample stimuli; in the other, the trial outcomes were nondifferentially associated with the sample stimuli. The key result was that delayed discrimination performance was superior with differential trial outcomes than with nondifferential outcomes, this superiority increasing the longer the sample-test retention interval.

This pattern of findings suggests that differential reinforcement affixes different expectancies to the sample stimuli; these expectancies in turn provide an extra source of information at the time of testing not available to subjects trained with nondifferential reinforcement. Furthermore, if the reinforcement expectancies elicited by the sample stimuli persist longer than the afferent traces of the sample stimuli, one can plausibly account for the short-term memory functions of differential subjects falling more slowly than those of nondifferential subjects; the former can take advantage of the longer lasting expectancies, whereas the latter cannot. A particularly clear instance of this effect was reported by DeLong and Wasserman (1981) (Figure 4.3).[3] Here, the differential outcomes involved different probabilities (0.2 and 1.0) of food

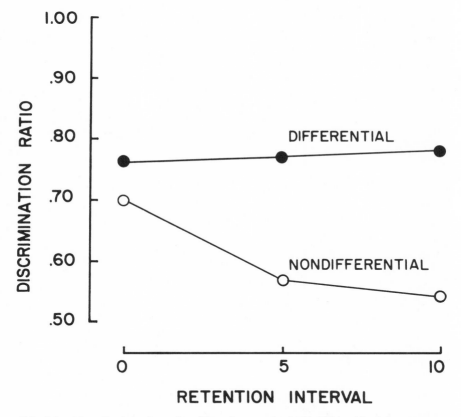

FIG. 4.3.   Mean discrimination ratios of four pigeons trained with differential reinforcement outcomes and four pigeons trained with nondifferential reinforcement outcomes after two different sample stimuli.

---

[3]The data come from Experiment 1B of DeLong and Wasserman (1981), and are the average of the last 30 of 120 days of training. Retention intervals were 0, 5, and 10 s long.

reinforcement affiliated with the two sample stimuli, whereas the nondifferential outcomes involved equal probabilities (0.6 and 0.6) of reinforcement.

## FURTHER ISSUES

Earlier, it was argued that the results of Honig and Wasserman (1981) could not be accounted for by assuming retrospective mediation of both delayed simple and conditional discrimination performance. The diverging memory functions (Figures 4.1 and 4.2) obtained with equally discriminable and rememberable sample stimuli strongly supported the involvement of prospective mediation.

Proceeding from Konorski's (1967) initial logic, then, we might assume that birds given only a delayed simple discrimination solve the problem prospectively, whereas birds given only a delayed conditional discrimination solve the problem retrospectively. Furthermore, since some of the birds in the Honig and Wasserman (1981) study (indeed, those whose data are depicted in Figure 4.1) received *both* problems intermixed in every session (as did those in our follow-up; see Figure 4.2), we might also assume that the very same subject can selectively adopt different methods of mediation depending upon the task demands. But would such assumptions be warranted? Perhaps not.

If we further contemplate how birds in the Honig and Wasserman (1981) investigation might master delayed simple and conditional discriminations, we can quite reasonably imagine that *both* problems could be solved prospectively. This account would follow from the notion that sample stimuli are capable of instigating retrieval of response rules, sets, or dispositions from long-term memory. Specifically, at the time of the sample stimulus, the pigeon encodes the appropriate response decision and remembers it. In the delayed simple discrimination, the bird's decision is merely to respond or not to respond, regardless of which test stimulus appears. In the delayed conditional discrimination, the decision has to include a stimulus element as well, e.g., for one sample stimulus, respond if the test stimulus is a vertical line but do not respond if it is a horizontal line; for the other sample stimulus, apply the opposite rule. Such a complex, stimulus-contingent decision could well be more difficult to remember over the course of the sample-test retention interval, and this might account for the different slopes of the memory functions obtained in delayed simple and delayed conditional discriminations.

To summarize: Konorski (1967) might assume prospective mediation of delayed simple discriminations and retrospective mediation of delayed conditional discriminations, whereas Honig and Wasserman (1981) have suggested the possibility that both discriminations entail prospective mediation. Clearly, each account expects prospective mediation of delayed simple discriminations. The accounts differ in the nature of the memory process mediating delayed conditional discriminations. Can we say whether delayed conditional discriminations are mediated prospectively or retrospectively?

## Confusion Errors

A few studies have examined delayed conditional discriminations in such a way as to yield evidence on prospective versus retrospective mediation. The evidence in two of these studies was in the form of confusion errors, which provide more than the usual amount of information concerning the basis of discriminative responding (also see Riley, Cook, & Lamb, 1981). In one investigation, Wasserman, Nelson, and Larew (1980) sought to determine if pigeons could discriminate and remember recent sequences of stimuli and responses. To this end, two-unit sequences of the form $(S_1-R_1)$, $(S_2-R_2)$ were used as samples in a delayed conditional go/no go discrimination procedure. With two sample side-key locations, four sample sequences resulted (Left-Left, Left-Right, Right-Left, and Right-Right). Each of these sequences was uniquely paired with reinforcement after only one of four different center-key test stimuli (Test 1, Test 2, Test 3, and Test 4, respectively); after the remaining sample-test sequences, nonreinforcement was the outcome. By monitoring the pigeon's rate of key pecking to the four test stimuli after the four sample sequences, it was possible to determine the degree to which the compound samples gained discriminative control over behavior at test.

Figure 4.4 portrays response rates to each of the four test stimuli after each of the four S-R sample sequences. Rates were highest on trials involving reinforcement; rates were lowest on nonreinforced trials in which the sample sequences ended in a different terminal unit than the corresponding reinforced trial sequences; and rates were intermediate on nonreinforced trials in which the sample sequences ended in the same terminal unit as the corresponding reinforced trial sequences. If the rate of response on nonreinforced trials is inversely related to the bird's confidence in identifying incorrect sequences, then the data tell us that confusions among the various sample sequences occurred most frequently when the last S-R unit of two different samples was the same; the left-left and right-left samples were quite likely to be confused with one another as were the left-right and right-right samples. Such a confusion pattern better reflects the relative recency of past S-R sample units than any differences in expected response options. Thus, these data support retrospective processing of two-unit S-R sequences.

In the second investigation, Roitblat (1980) employed a delayed symbolic matching-to-sample procedure. Color sample stimuli were followed by line-orientation test stimuli or vice versa. Three samples and three choice stimuli were used, each trial involving one sample and three choice stimuli. Roitblat selected the task stimuli so that the two most similar choice stimuli corresponded with the two most dissimilar sample stimuli, and the two most dissimilar choice stimuli corresponded with the two most similar sample stimuli. If the bird was retaining attributes of the sample stimulus, then with increasing retention intervals, similar samples would be more likely to be confused than would dissimilar samples. Alternately, if the bird was anticipat-

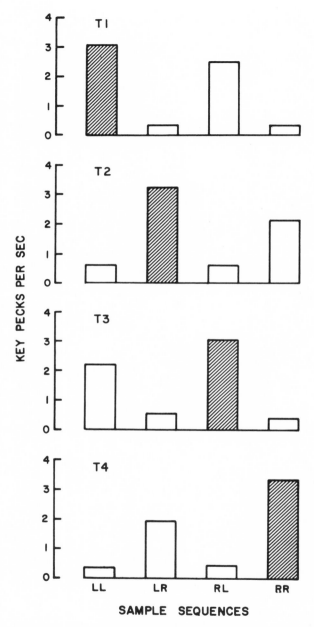

FIG. 4.4.    Mean rates of key pecking by six pigeons to each of four test stimuli (T1, T2, T3, and T4) after each of the four sample sequences (LL, LR, RL, and RR) on selected 5-day blocks of training in Experiment 1 of Wasserman et al. (1980). Only LLT1, LRT2, RLT3, and RRT4 trials involved food reinforcement (shaded bars); the remaining trials involved nonreinforcement (open bars).

ing the correct test stimulus, then with increasing retention intervals, confusions between similar test stimuli should increase more rapidly than confusions between dissimilar test stimuli.

The findings showed that, with increasing delays between sample and choice stimulus presentations, the pigeons were more likely to confuse test stimuli than to confuse sample stimuli. Such a pattern of results is consistent with prospective mediation, at least to the degree that Roitblat's (1980) metric of confusion is an unbiased measure (for more on potential problems with the calculation of discriminal distances, see Roberts, 1982 and Roitblat, 1982).

We are thus led to conclude that pigeons may solve delayed conditional discriminations either prospectively (as in the case of Roitblat, 1980) or retrospectively (as in the case of Wasserman et al., 1980). Aside from the fact that more response decisions were probably involved in the task of Wasserman et al. than in that of Roitblat, there is no obvious reason for pigeons adopting different mediational methods in performing the two discriminations.

## Retention Interval Stimuli

Another way to address the nature of the mediational process involved in delayed-discrimination performance is to vary the events that intervene between sample stimulus offset and test stimulus presentation. One study along these lines was conducted by Wasserman and DeLong (1981). The tactic here was to supplement standard go/no go matching-to-sample training (shown in the top portion of Table 4.1) with possible memory aids of two types: one that should improve performance according to retrospective mediation and one that should improve performance via prospective mediation. To the degree that these aids actually facilitate memory performance, we would have supportive evidence for these two kinds of mediation.

Two groups of four pigeons were investigated. In addition to the control trials shown in the top portion of Table 4.1, Group S received training with retention-interval stimuli that were redundant to the prior sample stimuli, as shown in the middle portion of Table 4.1. Retrospection should here lead to better performance on redundant sample trials than on trials with no retention interval stimulation. On redundant sample trials, strong memory traces of Colors 3 and 4 can control test performance; however, on trials with no retention-interval stimulation, the rather older and weaker memory traces of Colors 1 and 2 would have to control test responding.

In addition to the control trials shown in the top portion of Table 4.1, Group T received training with retention interval stimuli that were redundant to the upcoming test stimuli, as shown in the bottom portion of Table 4.1. Prospection should here lead to better performance on redundant test trials than on trials with no interpolated stimuli. On redundant test trials, presentation of Colors 3 and 4 can permit the pigeon to narrow its test options either "to peck"

TABLE 4.1
Go/No Go Matching-to-Sample Procedures of Wasserman
and DeLong (1981)

| Trial Sequence | Trial Possibilities | | | |
|---|---|---|---|---|
| | 1 | 2 | 3 | 4 |
| Control Procedure | | | | |
| Sample Stimulus (5 s) | Color 1 | Color 1 | Color 2 | Color 2 |
| Retention Interval | | | | |
| (1, 5, or 10 s) | ——— | ——— | ——— | ——— |
| Test Stimulus (5 s) | Color 1 | Color 2 | Color 1 | Color 2 |
| Trial Outcome (2.5 s) | Food | Blackout | Blackout | Food |
| Intertrial Interval (20 s) | ——— | ——— | ——— | ——— |
| Redundant Sample Procedure | | | | |
| Sample Stimulus (5 s) | Color 1 | Color 1 | Color 2 | Color 2 |
| Retention Interval | | | | |
| (1, 5, or 10 s) | Color 3 | Color 3 | Color 4 | Color 4 |
| Test Stimulus (5 s) | Color 1 | Color 2 | Color 1 | Color 2 |
| Trial Outcome (2.5 s) | Food | Blackout | Blackout | Food |
| Intertrial Interval (20 s) | ——— | ——— | ——— | ——— |
| Redundant Test Procedure | | | | |
| Sample Stimulus (5 s) | Color 1 | Color 1 | Color 2 | Color 2 |
| Retention Interval | | | | |
| (1, 5, or 10 s) | Color 3 | Color 4 | Color 3 | Color 4 |
| Test Stimulus (5 s) | Color 1 | Color 2 | Color 1 | Color 2 |
| Trial Outcome (2.5 s) | Food | Blackout | Blackout | Food |
| Intertrial Interval (20 s) | ——— | ——— | ——— | ——— |

or "not to peck" prior to presentation of the test Colors 1 and 2; however, on trials with no interpolated stimuli, the effective response codes cannot be narrowed either "to peck" or "not to peck" until the Color 1 or Color 2 test stimuli are actually presented.

Training under these procedures lasted 68 days. For the first 64 days, the scheduled retention intervals were 1 s and 5 s; for the final 4 days, the retention intervals were 1 s and 10 s. Colors 1 and 2 were red and green; Colors 3 and 4 were yellow and purple. Daily sessions comprised 64 trials, a random half being control trials involving no retention-interval stimuli and the other half being experimental trials involving stimuli interpolated into the retention interval.

As is typical in this conditioning situation, strong discriminative performance emerged on control trials in 2 weeks, such discrimination being shown by high rates of test responding on matching trials and low rates of test responding on nonmatching trials (see also Figure 4.4). A simple ratio expresses this discrimination: the rate of response on matching trials divided by the sum of the response rates on positive and negative trials (see also Figures 4.1 and 4.3). As is also typical of performance in this situation, control trial discrimination ratios were higher following a short retention interval than following a long retention interval.

Figure 4.5 documents this forgetting effect in Groups S and T and also shows performance on experimental trials. The data in this figure come from the final 8 days of training, and thus represent terminal performance. The top left panel shows the data of Group S. Here, experimental trial performance exceeded control trial performance. Further, at longer retention intervals,

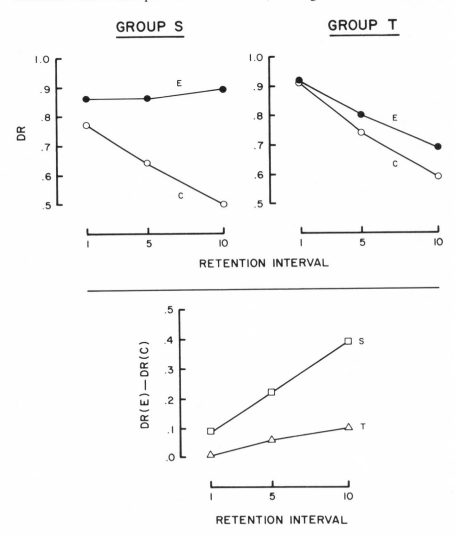

FIG. 4.5.   Mean discrimination ratios (top) and discrimination difference scores (bottom) of Group S ($n = 4$) and Group T ($n = 4$) of Wasserman and DeLong (1981).

control trial performance fell, whereas experimental trial performance did not. Thus, these results attest to the potency of redundant sample stimulation as an aid to matching-to-sample performance. The top right panel shows the data of Group T. Here, there was no reliable facilitation in performance from redundant test stimulation, both experimental and control trial performance falling at about the same rate as the retention interval was increased. To compare the extent to which retention interval stimulation improved discriminative performance in Groups S and T, experimental-control difference scores were computed. These are shown in the bottom portion of Figure 4.5, and reveal much greater improvement in Group S than Group T. The marked facilitation in memory performance occasioned by redundant sample stimulation coupled with the absence of notable facilitation by redundant test stimulation must then be construed as evidence in support of retrospective mediation and in opposition to prospective mediation.

Yet another technique for studying prospective and retrospective mediation involves retention-interval stimulation (and sample stimulation also). Stonebraker (1981; also see Nevin & Liebold, 1966) initially trained pigeons on two delayed conditional discriminations: matching-to-identical samples and symbolic matching-to-sample. This training entailed one set of sample stimuli (red and green) and two sets of choice stimuli (red-green and vertical line-horizontal line). In addition to these task stimuli, so-called instructional stimuli (circle and triangle) were superimposed on the sample stimuli and continued into the retention interval; these stimuli signalled whether the choice stimuli on that trial were to involve colors (identity task) or lines (symbolic task).

According to a retrospective account of animal memory, traces of sample stimulation serve as the functional cues for accurate choice-test performance (Roberts & Grant, 1976); because the samples for identity and symbolic problems come from the same set, there would be no basis for expecting the instructional stimuli to gain control of sample-stimulus processing. However, according to a prospective account, the instructional stimuli should control the subject's test-response dispositions; the instructional stimuli enable the bird to anticipate the upcoming test stimuli and thus narrow the range of possible choice behaviors.

To test whether the instructional stimuli did in fact influence sample-stimulus processing, Stonebraker (1981) occasionally miscued the test stimuli that were actually presented (e.g., he cued an identity test, but gave a symbolic test). Performance on miscued trials was notably inferior to performance on correctly cued trials. This result is thus consistent with a prediction of the prospective account: Pigeons can learn to anticipate the test options they will be given. The result does not tell us whether prospection is the normal mode of mediation in delayed conditional discriminations. Nor is the result easily distinguished from stimulus generalization decrement, miscued tests involving a change from the normal pattern of stimulation on correctly cued trials.

As was the case with research into confusion errors, we again find support for mediation by both prospection (Stonebraker, 1981) and retrospection (Wasserman & DeLong, 1981) in delayed conditional discriminations. The source of this inconsistency is not readily apparent.

## Other Tactics

Still other studies have either directly or indirectly pursued prospective and retrospective mediation of animal memory performance. Grant (1982b) used a proactive facilitation paradigm and a delayed conditional choice procedure to this end. Samples of color (red and green), number of key pecks (one and 20), and the occurrence or nonoccurrence of a reinforcer (food and no food) were employed. The choice stimuli were always red and green keys. Red was the correct choice given a prior sample of red, 20 pecks, or food; green was the correct choice given a prior sample of green, one peck, or no food.

On each trial, either one, two, or three sample presentations were given. With multiple samples, half of the trials involved the successive presentation of the same event; the other half of the trials involved the successive presentation of samples from different event classes. A prospective view of memory predicts equivalent levels of choice accuracy on same and different trials. The logic here is that the main role for the sample stimuli is to retrieve response rules that serve to guide the pigeon toward the correct choice stimulus. To the degree that each of a set of three samples sets a response disposition that is equally effective when given alone, giving two or more samples prior to the choice test should progressively add to the accuracy of test performance, whether the same sample is repeated or two or more different samples signifying the same rule are given. A retrospective view, on the other hand, predicts higher levels of choice accuracy on same trials than on different trials. The rationale here is that successive presentations of the same sample should strengthen the memory trace of that sample, whereas the successive presentation of different samples should establish multiple traces, each rising to a strength level equal to that accruing on only a single presentation. Because choice responding is assumed by trace theorists (e.g., Roberts & Grant, 1976) to be controlled by the strongest active memory trace, accuracy should be higher on same sample trials than on different sample trials.

Grant (1982b) found that the accuracy of choice responding was independent of whether successive sample presentations within a trial entailed the same or different, but associatively equivalent samples. He thus concluded that physically different samples which are associated with selection of the same choice stimulus initiate the same prospective process, in the form of an instruction for test responding.

Grant (1982a) pursued this analysis, but now with a proactive interference paradigm. On control trials in a delayed conditional choice procedure, samples of number of pecks (one and 20) and the occurrence or nonoccurrence of a reinforcer (food and no food) were employed. The choice stimuli were always red and green keys. Red was the correct choice given a prior sample of 20 pecks or food; green was the correct choice given a prior sample of one peck or no food. On interference trials, a presample was given that was associated with the incorrect choice stimulus. Half of the presamples were from the same dimension as the sample and half were from the other dimension.

Grant (1982a) found that the amount of proactive interference was independent of whether the presample and the sample used on a given trial were selected from the same dimension or from different dimensions. He again offered that these results are in line with a prospective or instructional encoding hypothesis of animal memory, antagonistic response rules competing with one another regardless of which samples happened to retrieve those rules.

Honig and Dodd (1983) also reported data consonant with prospective mediation. In their delayed go/no go discrimination procedures, trials always began with a red or a green stimulus and, after a delay, ended with a blue stimulus. Pecking the blue stimulus yielded reinforcement or nonreinforcement, depending upon the color of the key that came on at the start of the trial and the orientation of a line stimulus that either accompanied the initial color or the terminal blue stimulus. Were prospective mediation involved in discriminative performance, then we would expect that test performance should be better when the line stimulus joined the initial color than when it joined the terminal stimulus; in the former case response decisions could be narrowed before onset of the blue stimulus, whereas in the latter case no such early narrowing could occur (also see Wasserman & DeLong, 1981). This was indeed what Honig and Dodd found, particularly when the delay between the initial color and the terminal stimulus was lengthened. As was the case in the study by Honig and Wasserman (1981), some involvement of prospective mediation seems likely. But, as was also the case in the Honig and Wasserman study, it is not entirely clear whether one or both of the tasks that were investigated entailed prospective processing.

Although not directly aimed at the prospection/retrospection issue, a study by Wallace, Steinert, Scobie, and Spear (1980) yielded data consistent with retrospective mediation of successive matching-to-sample performance by rats. When lever pressing produced food on trials with the same sample and test stimuli but did not produce food on trials with different sample and test stimuli, rats' loss of discriminative responding was faster for visual than for auditory cues. This result is plausibly interpreted in terms of different rates of trace decay for visual than for auditory sample stimuli, a retrospective account. However, the result is not consistent with a prospective interpretation, where the complexity of any anticipatory response decisions ought to be the same no matter what the task stimuli.

The study of Wallace et al. (1980) and constructive replications of it with other species (e.g., Nelson & Wasserman, 1981) bear an interesting relation-

4. PROSPECTION AND RETROSPECTION

ship to the work of Honig and Wasserman (1981). In the Honig and Wasserman investigation, identical sample stimuli were used; differently sloped memory functions under different experimental conditions therefore suggested the involvement of prospective mediation. In the Wallace et al. project, on the other hand, different sample stimuli were employed; differently sloped memory functions under otherwise identical experimental circumstances thus implied the operation of retrospective mediation. The force of each of these conclusions, of course, rests upon exclusionary logic. We should be very cautious in deciding the issue either way, for subjects may be using *both* prospection and retrospection in any given problem, but to varying degrees, and possibly at different times.

Finally, although not inspired by the prospection/retrospection issue nor even involving associative learning procedures, the results of a study by Whitlow (1975) merit consideration here. Whitlow presented rabbits with pairs of successive 1-s tone stimuli selected from a set of two (530 and 4000 Hz); 30, 60, or 150 s separated the first and second tones of each pair, and 150 s separated the second tone of one pair from the first tone of the next. Tone pairs were scheduled in such a way that a random half entailed the same auditory frequency and the other half involved the two different auditory frequencies. At the shortest (30-s) interstimulus interval, the rabbits' orienting response to the second tone was much smaller when it matched the first stimulus than when it did not. As the interstimulus interval was lengthened to 150 s, responding to the second matching tone rose until it no longer differed from the high level of responding observed to a second nonmatching tone.

How do these results relate to processes of animal memory? First, although the measured response of peripheral vasoconstriction was unconditionally evoked by the auditory stimuli themselves and was not a conditioned response of the sort with which we have been dealing, the method that Whitlow (1975) used was basically Konorski's (1959) successive matching-to-sample procedure. Second, as in other go/no go studies using Konorski's procedure, forgetting was asymmetrically manifested, here as a rise in responding to matching test stimuli, with little systematic change in responding to nonmatching test stimuli (c.f., Nelson & Wasserman, 1978).

Third, the results seem to be harmoniously embraced by a retrospective theory of memory. The first stimulus of a pair lays down a sensory trace that decays until onset of the second stimulus. If comparison of the second stimulus with the trace of the first yields a "match," a small orienting response is evoked; if, on the other hand, the comparison fails to yield a "match," a large orienting response is elicited. Thus, lengthening the interstimulus interval between matching auditory stimuli will lead decreasingly to a comparison response of "match," and to an accompanying rise in vasoconstriction. However, lengthening the interstimulus interval between nonmatching auditory stimuli will have little effect upon the amplitude of vasoconstriction responses; the rabbit should manifest large vasoconstriction responses whether it correctly remembers the prior auditory stimulus (and makes a "nonmatch" comparison response) or it completely forgets that another auditory stimulus was even presented (and treats the second as a novel or "first" stimulus).

But fourth, even a prospective account can make sense of these data. Although admittedly speculative, if one assumes that hearing a tone leads the animal to anticipate another presentation of the same stimulus, despite the fact that matching and nonmatching trials are equiprobable, then expected test stimuli should support smaller orienting responses than surprising test stimuli (Wagner, Rudy, & Whitlow, 1973). Lengthening the interstimulus interval will decrement discriminative test performance, here because the clarity or distinctiveness of the mediating expectancy may wane over time. And the change in responsivity will be most evident on matching trials, because the surprise value of the second tone will rise the longer it has been since the first stimulus was presented; on nonmatching trials, the second tone will be surprising and trigger strong orienting responses whether the animal effectively remembers the first tone or forgets it.

Whitlow's (1975) work thus reveals that the processes of animal memory can be productively investigated with associative and nonassociative learning procedures. Memory evidently cuts across both associative and nonassociative learning. The task of distinguishing prospective from retrospective mediation is, however, no less formidable with nonassociative learning techniques.

## CONCLUSION

The central concern of this chapter is with Konorski's (1967) distinction between prospection and retrospection as processes mediating delayed discrimination performance. Whereas prospection emphasizes expected outcomes or action patterns as mediators of discriminative test behavior, retrospection stresses perseverative sensory processes that are instigated by prior external stimuli.

Although most theorists (e.g., Roberts & Grant, 1976) have proposed retrospective accounts of animal memory, prospection may play a much greater role than is usually imagined (see also Honig & Thompson, 1982). An examination of recent evidence, while not definitive, suggests that even complex delayed discriminations may be mastered on the basis of anticipated response rules or reinforcement outcomes. What is not yet known is whether a given discrimination will be selectively solved by prospective or retrospective mediation. Nor is it known whether a given problem is capable of being solved by more than one mediational strategy. Pursuit of these and other questions cannot help but provide us with further evidence on the validity of the prospection/retrospection distinction.

The assumption that clear tests can be devised that will divulge the operation of prospection and retrospection in animal short-term memory procedures has been implicit in this chapter. Without such clear tests, we would have to conclude that the prospection/retrospection dichotomy is hopelessly hypothetical. Several methods have been described which appear to have promise in disclosing the selective involvement of prospection and retrospection. More tests will have to be devised, however, if we are to gain great confidence in the distinction. Whatever the fate of Konorski's (1967) proposal, its experimental examination will continue to disclose new empirical facts about a long neglected field of inquiry: animal memory.

Finally, brief mention can be made of the place of prospection and retrospection in what has been called the stream of consciousness (see also Honig, 1981 for a more modern treatment of the problem). Writing on this matter in 1894, C. Lloyd Morgan (1894/1896) observed that: "Our past life, which we can review in memory, is an extension backwards through retrospective thought of experience gained in the present moment of consciousness. Our anticipations of the future are a similar extension forwards of this experience. Anticipation is prospective representation [p.113]."

Charles S. Sherrington (1906) objectified these mentalistic ideas and related them to the evolution of brain and intelligence: "It is the long serial reactions of the 'distance receptors' that allow most scope for the selection of those brute organisms that are fittest for survival in respect to elements of mind [p. 333]." Processes mediating the time since a stimulus and a completed behavioral act often entail a cognitive component:

> In the time run through by a course of action focussed upon a final consummatory event, opportunity is given for instinct, with its germ of memory however rudimentary and its germ of anticipation however slight, to evolve under selection, that mental extension of the present backward into the past and forward into the future which in the highest animals forms the prerogative of more developed mind [p. 332].

Thus, the processes of prospection and retrospection, although most clearly revealed in seemingly unnatural laboratory tasks, may participate in such vital patterns of action as finding food, courting mates, and fleeing foes. Future study of memory processes may very well shed further light on the ecology and evolution of complex cognitive capabilities.

## REFERENCES

Blough, D. S. Delayed matching in the pigeon. *Journal of the Experimental Analysis of Behavior*, 1959, *2*, 151–160.

Bottjer, S. W., & Hearst, E. Food delivery as a conditional stimulus: Feature-learning and memory in pigeons. *Journal of the Experimental Analysis of Behavior*, 1979, *31*, 189–207.

Brodigan, D. L., & Peterson, G. B. Two-choice conditional discrimination performance of pigeons as a function of reward expectancy, prechoice delay, and domesticity. *Animal Learning and Behavior,* 1976, *4,* 121–124.

Cumming, W. W., & Berryman, R. The complex discriminated operant: Studies of matching-to-sample and related problems. In D. I. Mostofsky (Ed.), *Stimulus generalization.* Stanford, CA: Stanford University Press, 1965.

DeLong, R. E., & Wasserman, E. A. Effects of differential reinforcement expectancies on successive matching-to-sample performance in pigeons. *Journal of Experimental Psychology: Animal Behavior Processes,* 1981, *7,* 394–412.

Fletcher, H. J. The delayed response problem. In A. M. Schrier, H. F. Harlow, & F. Stollnitz (Eds.), *Behavior of nonhuman primates.* New York: Academic Press, 1965.

Grant, D. S. Intratrial proactive interference in pigeon short-term memory: Manipulation of stimulus dimension and dimensional similarity. *Learning and Motivation,* 1982, *13,* 417–433. (a)

Grant, D. S. Prospective versus retrospective coding of samples of stimuli, responses, and reinforcers in delayed matching with pigeons. *Learning and Motivation,* 1982, *13,* 265–280. (b)

Honig, W. K. Working memory and the temporal map. In N. E. Spear & R. R. Miller (Eds.), *Information processing in animals: Memory mechanisms.* Hillsdale, NJ: Lawrence Erlbaum Associates, 1981.

Honig, W. K., & Dodd, P. W. D. Delayed discriminations in the pigeon: The role of within-trial location of conditional cues. *Animal Learning and Behavior,* 1983, *11,* 1–9.

Honig, W. K., & Thompson, R. K. R. Retrospective and prospective processing in animal working memory. In G. H. Bower (Ed.), *The psychology of learning and motivation.* New York: Academic Press, 1982.

Honig, W. K., & Wasserman, E. A. Performance of pigeons on delayed simple and conditional discriminations under equivalent training procedures. *Learning and Motivation,* 1981, *12,* 149–170.

Hunter, W. S. The delayed reaction in animals and children. *Behavior Monographs,* 1913, *2,* 1–86.

Konorski, J. A. A new method of physiological investigation of recent memory in animals. *Bulletin de l'Academie Polonaise des Sciences Serie des Sciences Biologiques,* 1959, *7,* 115–117.

Konorski, J. A. *Integrative activity of the brain.* Chicago, IL: University of Chicago Press, 1967.

Konorski, J. A., & Lawicka, W. Physiological mechanism of delayed reactions: I. The analysis and classification of delayed reactions. *Acta Biologiae Experimentalis,* 1959, *19,* 175–197.

Morgan, C. L. *An introduction to comparative psychology.* London: Scott, 1896. (Originally published, 1894.)

Nelson, K. R., & Wasserman, E. A. Temporal factors influencing the pigeon's successive matching-to-sample performance: Sample duration, intertrial interval, and retention interval. *Journal of the Experimental Analysis of Behavior,* 1978, *30,* 153–162.

Nelson, K. R., & Wasserman, E. A. Stimulus asymmetry in the pigeon's successive matching-to-sample performance. *Bulletin of the Psychonomic Society,* 1981, *18,* 343–346.

Nevin, J. A., & Leibold, K. Stimulus control of matching and oddity in a pigeon. *Psychonomic Science,* 1966, *5,* 351–352.

Pavlov, I. P. *Conditioned reflexes.* Oxford: Oxford University Press, 1927.

Peterson, G. B., & Trapold, M. A. Effects of altering outcome expectancies on pigeons' conditional discrimination performance. *Learning and Motivation,* 1980, *11,* 267–288.

Peterson, G. B., Wheeler, R. L., & Armstrong, G. D. Expectancies as mediators in the differential-reward conditional discrimination performance of pigeons. *Animal Learning and Behavior,* 1978, *6,* 279–285.

Peterson, G. B., Wheeler, R. L., & Trapold, M. A. Enhancement of pigeons' conditional discrimination performance by expectancies of reinforcement and nonreinforcement. *Animal Learning and Behavior*, 1980, *8*, 22–30.

Riley, D. A., Cook, R. G., & Lamb, M. R. A classification and analysis of short-term retention codes in pigeons. In G. H. Bower (Ed.), *The psychology of learning and motivation*. New York: Academic Press, 1981.

Roberts, W. A. Premature closure of controversial issues concerning animal memory representations. *The Behavioral and Brain Sciences*, 1982, *5*, 384–385.

Roberts, W. A., & Grant, D. S. Studies of short-term memory in the pigeon using the delayed matching-to-sample procedure. In D. L. Medin, W. A. Roberts, & R. T. Davis (Ed.), *Processes of animal memory*. Hillsdale, NJ: Lawrence Erlbaum Associates, 1976.

Roitblat, H. L. Codes and coding processes in pigeon short-term memory. *Animal Learning and Behavior*, 1980, *8*, 341–351.

Roitblat, H. L. Representations and cognition. *The Behavioral and Brain Sciences*, 1982, *5*, 394–401.

Sherrington, C. S. *The integrative action of the nervous system*. New Haven, CT: Yale University Press, 1906.

Smith, L. Delayed discrimination and delayed matching in pigeons. *Journal of the Experimental Analysis of Behavior*, 1967, *10*, 529–533.

Stonebraker, T. B. *Retrospective versus prospective processes in delayed matching-to-sample*. Unpublished doctoral dissertation, Michigan State University, 1981.

Terry, W. S., & Wagner, A. R. Short-term memory for "surprising" versus "expected" unconditioned stimuli in Pavlovian conditioning. *Journal of Experimental Psychology: Animal Behavior Processes*, 1975, *104*, 122–133.

Trapold, M. A. Are expectancies based upon different positive reinforcing events discriminably different? *Learning and Motivation*, 1970, *1*, 129–140.

Wagner, A. R., Rudy, J. W., & Whitlow, J. W. Rehearsal in animal conditioning. *Journal of Experimental Psychology*, 1973, *97*, 407–426.

Wallace, J., Steinert, P. A., Scobie, S. R., & Spear, N. E. Stimulus modality and short-term memory in rats. *Animal Learning and Behavior*, 1980, *8*, 10–16.

Wasserman, E. A. Successive matching-to-sample in the pigeon: Variations on a theme by Konorski. *Behavior Research Methods and Instrumentation*, 1976, *8*, 278–282.

Wasserman, E. A. Comparative psychology returns: A review of Hulse, Fowler, and Honig's *Cognitive processes in animal behavior*. *Journal of the Experimental Analysis of Behavior*, 1981, *35*, 243–257.

Wasserman, E. A. Further remarks on the role of cognition in the comparative analysis of behavior. *Journal of the Experimental Analysis of Behavior*, 1982, *38*, 211–216.

Wasserman, E. A. Is cognitive psychology behavioral? *Psychological Record*, 1983, *33*, 6–11.

Wasserman, E. A., & DeLong, R. E. *Prospection and retrospection in animal memory*. Paper presented at the annual meeting of the American Psychological Association, Los Angeles, August 1981.

Wasserman, E. A., Grosch, J., & Nevin, J. A. Effects of signalled retention intervals on pigeon short-term memory. *Animal Learning and Behavior*, 1982, *10*, 330–338.

Wasserman, E. A., Nelson, K. R., & Larew, M. B. Memory for sequences of stimuli and responses. *Journal of the Experimental Analysis of Behavior*, 1980, *34*, 49–59.

Whitlow, J. W. Short-term memory in habituation and dishabituation. *Journal of Experimental Psychology: Animal Behavior Processes*, 1975, *104*, 189–206.

# 5 Anticipation and Intention in Working Memory

Werner K. Honig
*Dalhousie University*

Peter W. D. Dodd
*St. Mary's University**

## INTRODUCTION

Each trial in an experiment on working memory in animals incorporates three essential events: an initial stimulus (IS) or "sample," a test stimulus (TS) or a pair of test stimuli, and a response to the test stimulus or stimuli. A memory interval (MI) usually separates the initial and test stimuli. In delayed-matching-to-sample (DMTS) and delayed-conditional-matching-to-sample (DCMS), the single IS is followed by a pair of test stimuli, and the accuracy of choice between the latter provides the index of discrimination and memory. Two less familiar prcedures employ only a single TS on each trial. These are the delayed simple discrimination (DSD) and the delayed conditional discrimination (DCD). In the DSD, the IS (such as a color) provides the necessary information regarding the trial outcome, and differences between the test stimuli (such as line orientations) are irrelevant. In the DCD, the outcome depends on a conditional relationship between the IS and the TS. For example, responding to one line orientation will be reinforced only if that TS follows one initial color, while responding to the other TS in reinforced only if it follows the other initial color. In all of these procedures, correct responding to the TS is cued by the IS that precedes it. To master a working memory problem, the animal must acquire the discrimination between the initial stimuli; to master DMTS, DCMS, and the DCD, it must also discriminate between the test stimuli. To respond correctly to the TS on a given trial in any procedure, the subject must remember the IS throughout the MI. The form of this memory is the principal topic of this chapter.

---

*Peter W.D. Dodd is now at Bell Northern Laboratories, Ottawa, Canada.

In recent theoretical analyses of these memory problems, emphasis has shifted from a passive, retrospective memory process initiated at the time of the TS to a more active, prospective process initiated at the time of the IS. Evidence for these processes is reviewed by Honig and Thompson (1982) and by Wasserman in Chapter 4 of this book. The concept of "prospection" implies that some aspect of the trial can be anticipated: the trial outcome, the correct response to be made, the stimulus to be chosen, the quality or quantity of the reinforcer, and so forth. This anticipation can direct or influence performance at the time of the TS. In contrast, the retrospective view implies that the IS is remembered through the MI and this memory provides the necessary information at the time of the TS.

The distinction between retrospection and prospection can perhaps be clarified by the concept of a "response instruction" generated in the course of a trial (Honig, 1978). In prospective remembering, such an instruction is generated early in the trial, is based on information available at that time, and is remembered through the MI. In retrospective remembering, such an instruction is not generated early in the trial, and responding to the TS is based on information from the IS that is still available in working memory. Current research, some of it reviewed here, has led to the conclusion that response instructions are better remembered than stimuli, and prospective remembering is therefore more "robust" than retrospective remembering; forgetting is slower in the former.

In spite of the evidence and argument in favor of this distinction, there has been little analysis of the nature of prospective working memory. Experimental efforts have been focused, by and large, on the comparison of forgetting functions obtained with different procedures which are likely to favor either retrospective or prospective processing. In this chapter, we will go beyond the original notion of the "response instruction," and consider the "content" of prospective working memory. We will discuss two of its aspects: the anticipation of events and conditions in a trial (particularly the trial outcome), and the intention of making a specific response. We will provide evidence for both of these aspects and we will also suggest that the former (anticipation) can cue or direct the latter (response intention) in the course of a trial.

Anticipations and intentions are detected only indirectly, and the art of research in the area of working memory is the design of experiments that would reveal them. However, not many experiments have been directly addressed to the issue, and our discussion will in large part involve the interpretation of data that have already been obtained in the context of other experimental and theoretical analyses.

## RESPONSE INTENTIONS

In most working-memory paradigms, the IS can in principle be described as instructive; it indicates to the subject what response it should make to the TS, after the MI, to obtain reward. However, in many studies the IS probably does not function in this manner, particularly when the subject would have to encode stimulus information regarding the TS into the instruction ("respond to the red key"). Forgetting is generally more rapid in such problems than it is in procedures where such stimulus information regarding the correct TS is not required; indeed, this comparison is usually taken to indicate that the subject uses a retrospective process in delayed matching and delayed conditional discrimination procedures.

Wasserman describes such comparisons in Chapter 4, beginning with a study by Smith (1967). On some trials, the color on a center key indicated to the pigeon which of two side keys would be correct following the MI. On other trials, the center key served as a sample for a DMTS procedure. Remembering was superior in the first condition, but Smith used only a limited range of MIs (up to 5 s). We may assume that the cue to turn to one or the other side key would generate a simple response intention which does not involve the color of the side key.

## Whalen's Experiments

A more extensive set of comparisons was carried out by Whalen (1979). He used traditional DMTS as the procedure which required a combination of information from the IS and the TS. Generally, the stimuli were red and green, presented both as samples and as comparisons. The alternate procedure was a delayed simple discrimination (DSD) in which the initial stimuli (which usually differed from the DMTS cues) indicated the pattern of responding to a single TS which would procure reinforcement. Following one IS, the schedule was fixed interval (FI) 10 s; following the other IS, the schedule was differential reinforcement of other behavior, or DRO 10 s. This required the pigeon not to peck at the TS for 10 s to obtain reinforcement. The color of the TS was blue on all trials in the DSD procedure.

In a number of systematic replications with these procedures, Whalen (1979) found that remembering over the MI was consistently better in the DSD than in DMTS. However, the comparison was problematic for several reasons. The performance measures differed; the DMTS yielded percent correct of choices within trials, while the DSD was assessed by comparing response rates to the TS across trials. Whalen adapted the rate measure to make it commensurate with proportion correct; the difference in performance between the DSD and the DMTS was maintained. In DMTS, the proportion of reinforced trials depends on accuracy of performance; when memory decreases as a function of the MI, so does the number of reinforced trials. On the other hand, each trial in the DSD was reinforced because the TS remained on until the subject met the response criterion. Finally, the DMTS was the more difficult procedure, and performance was usually poorer even with no memory requirement, or only a short one. This makes it hard to compare the slopes of forgetting functions.

Whalen (1979) overcame these difficulties to a large extent in an interesting experiment in which he manipulated a variable that was orthogonal to other aspects of the procedure. He varied the opportunity for the subjects to "practice" remembering in the two problems. He trained several groups of pigeons on the DMTS and the DSD with no MI. The initial stimuli were red and green in both problems. In the DMTS the test stimuli were also red and green; in the DSD the TS was blue. Preliminary training ensured good performance on both problems with no MI. Twenty-five sessions then provided different conditions of "memory training" for five groups of four Ss each. One group received 10 sessions of training on both problems with MIs averaging 5s, and then 15 sessions of training with a mean MI of 15 s (DMTS15, DSD15). A second group received the same memory training only on DMTS, while the DSD was further trained with no MI (DMTS15, DSD0). A third group received the opposite condition (DMTS0, DSD15). A fourth group received further training with no memory requirement (DMTS0, DSD0). A fifth group was given memory training on both problems, but only on half of the trials (DMTS15/0, DSD15/ 0). All birds were then tested in two sessions with 15-s MIs in all of the DMTS and DSD trials.

The results from the test sessions are shown in Table 5.1. On the DSD trials, the subjects from the different groups performed at roughly the same level. Their discrimination ratios (DR), computed on the proportion of correct responses, ranged from .69 to .76. On DMTS trials, the groups that had no memory training performed only slightly above chance, while the remaining groups did significantly better, with ratios similar to those obtained from the DSD trials.

This finding argues strongly for different, independent memory processes, even though the same initial stimuli were used in both procedures. The pigeons seem predisposed to anticipate a response requirement and to generate and maintain a response intention. They cannot remember the IS for the purpose of choosing in the DMTS without special training. This result has some interesting implications. First, the memory for the response intention in the DSD does not seem to require or contain information about the IS. If it did, then

TABLE 5.1
Discrimination Ratios Obtained from Test Sessions in Whalen's (1979)
Experiment on Memory Training

| Group | Memory Task | |
| --- | --- | --- |
| | *DSD* | *DMTS* |
| DMTS15, DSD15 | .71 | .63 |
| DMTS15/0, DSD15/0 | .68 | .70 |
| DMTS15, DSD0 | .70 | .64 |
| DMTS0, DSD15 | .75 | .54 |
| DMTS0, DSD0 | .76 | .55 |

performance on the DSD would have suffered without memory training, as it did in the DMTS. Second, the response intentions in the DSD do not seem to have mediated choice responding in DMTS. Remember that the same initial stimuli were used for both problems, and they presumably generated different response intentions in the DSD. It might be supposed that the intentions could also develop control over DMTS performance, as they could in principle serve as cues for the correct choice between stimuli, as well as the correct pattern of responding to a single TS. This does not seem to have been the case. The patterns of the "intended" responses in the DSD are very different from the single-choice response required on a DMTS trial, and this may have precluded any mediation. The finding is of particular interest because we can readily show that expectancies of differential outcomes can control choice or response rates in other conditional discriminations, even if response intentions do not.

## Delayed Simple and Conditional Discriminations

Data from delayed simple and the delayed conditional discriminations (DSD and DCD) have supported the distinction between prospective and retrospective memory processing. Wasserman reviews relevant research in Chapter 4; therefore, an experiment by Honig and Wasserman (1981) needs only brief description. In the DSD, an initial color provided information about availability of reinforcement for responding to the TS. The TS was a horizontal or vertical line orientation, but in the DSD this cue was irrelevant. In the DCD, reinforcement depended upon both color and line orientation, so that the subject could not develop a response intention at the time of the IS. Memory was clearly superior in the DSD, whether the comparison was carried out within subjects or between groups. (See Chapter 4 for further description, particularly Fig. 4.4; See also Fig. 5.4.)

The DSD is an easier problem for pigeons to acquire than the more complex DCD. This difference is confounded with the difference in paradigms that makes the comparison interesting. This problem was addressed by Honig and Dodd (1983), who used a conditional discrimination in two different arrangements. The DCD was the same as described above; trials began with red or green and ended with horizontal or vertical lines presented on a blue background. Only two color-line sequences (e.g., red-vertical and green-horizontal) yielded reinforcement. The conditional delayed discrimination (CDD) was arranged so that the colors and lines were combined as the initial stimuli, and the TS was a plain blue field. As in the DCD, only two color-line combinations signaled reinforcement. The same subjects learned both problems, and the same color-line pairings were positive in both.

After the subjects learned these problems, MIs were introduced; the distribution of MIs increased over blocks of sessions, although a 1-s MI was maintained in all conditions. Figure 5.1 summarizes the results in the form of

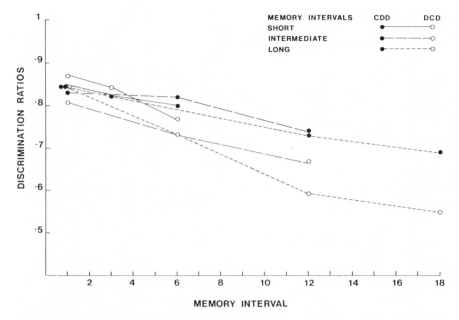

FIG. 5.1.   Forgetting of an initial stimulus over several MIs with pigeons trained on a delayed conditional discrimination (DCD) and on a conditional delayed discrimination (CDD) (Honig & Dodd, 1983).

forgetting curves. The DRs were very similar at a 1-s MI, but forgetting was faster in the DCD. This result is striking because the CDD was actually acquired more slowly than the DCD.

These studies suggest that, even when other factors are controlled, working memory is better when sufficient information is presented early in the trial, so that the subject can establish a response intention and remember it, rather than having to remember one item of information until the other becomes available. However, the pigeon could also have come to anticipate the reinforcement at the end of a positive trial, instead of, or in addition to the appropriate response. In Whalen's (1979) experiments, this problem did not arise, since all DSD trials ended in reinforcement. But in our experiments, (Honig & Dodd, 1983; Honig & Wasserman, 1981), the mechanism mediating working memory in the DSD is not clear. Was it a response intention, or did the expectations of reward and reward omission provide cues for responding to the test stimuli? This question leads us to differential outcome expectancies in working memory.

## ANTICIPATION OF TRIAL OUTCOMES

The differential outcome effect (DOE) is by now a familiar finding in animal discrimination learning and memory. Initial stimuli in a conditional discrimination signal not only which response is correct at the end of the trial, but also which of two positive outcomes will be presented. This is the differential outcome (DO) procedure. Acquisition is generally faster and retention of the IS over a working MI is generally better, with a DO procedure, than it is if the same positive outcome is used on all reinforced trials, or if the two outcomes are randomly presented with respect to the initial stimuli. Various differential outcomes have been used, including the kind of reward, such as food and water, the probability of reward, delay of reward, and so forth. The topic has been extensively reviewed elsewhere (Honig & Thompson, 1982; Peterson, 1984; Peterson & Trapold, 1980).

In general, writers agree upon a particular theoretical account of the DOE. They invoke differential outcome expectancies. Different expectancies of reinforcement or of some other trial characteristic are established in training, and one of these is activated on each trial by the IS. The expectancy is maintained through the MI. When the TS appears, the expectancy cues either a choice between two test stimuli, or the appropriate rate of responding to a single TS. Thus, the DOE is generally taken as evidence for prospective memory processing, as the expectancy is generated on each trial at the time of the IS.

The role of the memory of the IS is not clear in this interpretation of the DOE. The subject may simply forget the IS on each trial, as its role is taken over by the outcome expectancy. On the other hand, such an expectancy may enhance memory by making the initial stimuli more discriminable through differential associations. This question is certainly an important one for our account of the "content" of prospective memory. It is only resolved by analytic experimental strategies, and not by repeated demonstrations of the DOE. In one strategy, the expectancies generated by a set of initial stimuli are changed, and the effect is observed when the original discrimination or some variant of it is reinstated. In the other, differential outcome expectancies are trained with a new set of stimuli, and these replace the initial stimuli in a transfer procedure.

## Changes in Outcome Expectancies

Peterson and Trapold (1980) used the first approach. Only two groups from their rather complex experiment are described here. Pigeons were trained with a differential outcome procedure in which "true" matching of red was followed by food, and matching of green by a tone and by the advance of the program to the next trial. In the second phase of the study, these trial outcome expectancies were reversed "off baseline"; the new trial outcomes followed the initial stimuli directly. In the third phase, a conditional discrimination was reinstated. For one group,"true" matching to sample was required as before, but the trial outcomes were still reversed. A second group was rewarded for "mismatching," so that the contingencies as well as the outcomes were reversed.

Peterson and Trapold (1980) obtained remarkably good immediate transfer to the new contingencies (mismatching) in the group with the reversed outcomes. In contrast, performance with the original contingencies still in effect dropped well below 50% correct and recovered slowly. This suggests quite strongly that expectancies can govern the choice of the test stimuli, and that they overshadowed any residual direct control of responding by the initial stimuli.

## Transfer based on Differential Outcome Expectancies

Peterson (1984) studied the transfer of a discrimination of orthogonally related initial stimuli with an elegant factorial design. Four groups of pigeons learned a conditional matching procedure with red and green as initial stimuli and with horizontal and vertical lines as test stimuli. The outcomes, food or a tone, were differentially correlated with red and green for two groups, and nondifferentially presented for two others. The pigeons also received simple acquisition trials with a circle or a triangle on the key. Responding to these forms was directly reinforced with food or a tone, and these outcomes were again differentially associated with the circle and the triangle for two groups and randomly presented for two others.

In the transfer procedure, the forms replaced the colors as initial stimuli. Only one group had received DO procedures in both phases of training, and they were so arranged that the expectancies based on colors and forms would cue the same correct choices between the two test stimuli. This group showed immediate and almost complete transfer when the forms replaced the colors as initial stimuli. The other groups reverted to a chance level because they had no expectancies on which to base the transfer of control from colors to forms.

Edwards, Jagielo, Zentall, and Hogan (1982) used a transfer strategy that made "off-baseline" acquisition training unnecessary. Their pigeons were trained concurrently with two independent true-matching procedures; one involved colors as samples and as comparisons, and the other was based on geometric forms. Corn and wheat, as the trial outcomes, were differentially associated with the initial stimuli in both discriminations. Thus, for example, red or cross would precede corn, while green or circle would precede wheat. Transfer was introduced simply by exchanging colors and forms as the initial stimuli. This turned the problems into conditional matching procedures. The tranfer was arranged so that the newly defined correct choices were congruent with the differential outcome expectancies for half the subjects; this should result in positive transfer. The remaining birds were run on a negative transfer procedure, in which the new correct choices were incongruent with those that would be cued by the outcome expectancies. Edwards et al. observed some immediate transfer in the congruent procedure; the subjects started with 65% correct choices, and acquired the new discrimination more rapidly than did a control group with mixed outcomes. The incongruent group was slightly retarded in acquiring the second discrimination. Results can be seen in Figure 5.2. The degree of transfer was less than that obtained by Peterson (1984) and by Peterson and Trapold (1980). This can perhaps be explained by the discriminability of the outcomes. Different kinds of grain may not be as easily discriminated as reward versus no reward with advance to the next trial.

A recent study from our laboratory (Honig, Matheson, & Dodd, 1984) replicated and extended the transfer of discriminations based on outcome expectancies. We worked with the DCD, as this proved to be such a valuable tool in our prior research, and we also incorporated a negative transfer condition. Peterson (1984) did not use such a condition, and Edwards et al. (1982) obtained only a weak negative transfer effect. Furthermore, we maintained the original discrimination in the transfer phase. This should provide a sensitive method for assessing the role of outcome expectancies. Training with negative transfer would invalidate the original expectancies as cues in the original problem, and this change might well be reflected in a loss of the original discrimination.

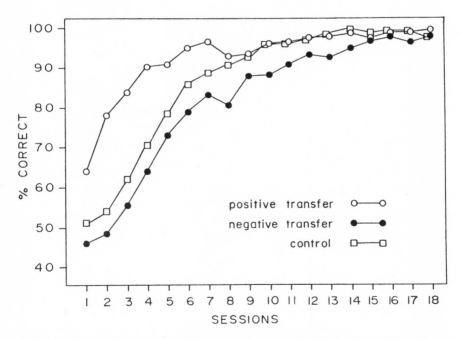

FIG. 5.2.   The transfer phase of a matching-to-sample problem in which two different kinds of grain reinforcer were cued by initial stimuli (Edwards et al., 1982).

Two groups of pigeons were first trained in a DCD with red and green as initial stimuli, and with grain and water as differential outcomes. Red-vertical sequences were rewarded with grain, and green-horizontal with water. The remaining sequences (red-horizontal and green-vertical) yielded no reward. (The outcomes and the positive sequences were counterbalanced across subjects.) When this discrimination was acquired, blue and white were introduced as initial stimuli. At first, they were associated off baseline with grain and water reward. The pigeons pecked at a cross-hatched black-and-white pattern immediately following blue to obtain the grain reward, and at the same pattern immediately following white to obtain the water reward. In the transfer phase, blue and white were used in the DCD as initial stimuli. Training was continued with the original red-green discrimination in alternate sessions.

For consistent transfer, outcomes were arranged so that the expectancies established off baseline would lead to correct performance. Thus, the blue-vertical sequence (like the red-vertical sequence) was followed by food, and the white-horizontal sequence (like the green-horizontal sequence) by water. In the inconsistent transfer condition, blue-horizontal was followed by food, and white-vertical by water. If outcome expectancies cue differential responding to the test stimuli, then the consistent procedure should lead to immediate transfer or facilitated acquisition, and the inconsistent procedure should lead to negative transfer or retarded acquisition.

Figure 5.3 provides the findings from the transfer phase. In the consistent procedure, performance on the blue-white problem was immediately very good; it was close to that achieved on the red-green discrimination during original training. The red-green discrimination was not affected by the introduction of the transfer problem. In the inconsistent transfer condition, performance on the blue-white problem started at or below "chance," and the red-green discrimination was disrupted. Eventually, the transfer discrimination was acquired, and the original discrimination approached its former level.

These finding support the notion of outcome expectancies as cues. They are initiated by the first stimulus, and they direct responding to the test stimulus. Unfortunately we did not introduce MIs in this procedure in order to test memory functions directly. Such functions were obtained in another, rather complex experiment using the DCD (Honig et al., 1984). The design was basically the same as in the experiment just described, except that the red-green and the blue-white DCDs were trained concurrently. The trial outcomes were arranged so that expectancies were consistent with correct response patterns to the test stimuli for one group, but inconsistent for another group. Acquisition was much better with the former procedure, so that subjects met performance criteria which allowed them to advance to significant memory intervals. Subjects for which outcome expectancies would cue inconsistent response patterns did not learn the problem well enough to be tested with memory intervals.

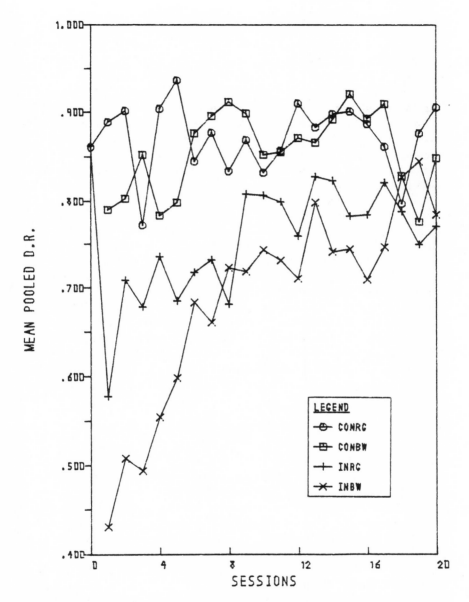

FIG. 5.3.  The transfer phase of a delayed conditional discrimination with differential food and water outcomes (Honig, Matheson, & Dodd, 1984).

## Discussion

These experiments indicate that outcome anticipations can be generated by initial stimuli, and that they support and even supplant such stimuli, or memories of such stimuli, in controlling differential choices or response rates to test stimuli. The transfer studies are particularly convincing, because the original stimuli are simply replaced by new ones that could not themselves control such differential behavior. When outcome expectancies were established off baseline that conflicted with those controlled by the original training stimuli, the latter lost their control over differential behavior. This suggests that the expectancies, although generated by the initial stimuli, eventually overshadowed them in controlling differential behavior to the test stimuli.

Outcome expectancies are obviously powerful in controlling behavior. This supports one possible interpretation, mentioned above, of the superior working memory observed in the DSD over the DCD (Honig & Wasserman, 1981), and in the CDD over the DCD (Honig & Dodd, 1983). The DSD and the CDD permit the subjects to develop anticipations of reward and nonreward on different trials, and these anticipations can then serve as cues for responding to the test stimuli. Support for this notion is provided by differential outcome experiments by Peterson and Trapold (1980), and by Peterson, Wheeler, and Trapold (1980). One positive outcome was reward, and the other was no reward with the opportunity to advance to the next trial. Better performance over an MI was obtained with this DO procedure than in a nondifferential condition, in which each correct choice was rewarded with grain. It appears, then, that anticipation of no reward can serve as a cue for discriminative responding.

## ANTICIPATION OF TRIAL CHARACTERISTICS

From the data reviewed so far, it is clear that animals can anticipate characteristics of the trial outcome: the quality, quantity, probability, or other features of the reinforcer. Such anticipations serve as cues for responding to test stimuli, and they have a profound effect on memory functions, as they would appear to replace retrospective memory of the IS. It is reasonable to suppose that animals can also learn to anticipate other features of the procedure within trials, features that are not differentially associated with the initial stimuli. Such features may remain constant over many trials. For example, the size of the reinforcer could be varied across blocks of sessions. The subject could adjust its behavior on a long-term basis. On the other hand, the trial characteristic may vary from trial to trial, being indicated by a differential cue on each trial. The cues which signal such trial characteristics would have to be independent of, and normally separate from, the initial stimuli. We will call such stimuli "informative," as they inform the subject on each trial regarding some significant value or characteristic that may affect behavior.

Performance will be affected by such general variations in trial characteristics. The subject may perform less well if it anticipates a small reward, even if it remembers the IS on a given trial. Thus, while changes in trial characteristics may appear to affect memory, this may actually be due to some other factor, such as the incentive to respond accurately. We do not know of any studies on the size of reward in working memory. But one treatment that has received attention is the signaled omission of the test stimuli, or "directed forgetting." This treatment generally produces poor remembering on "probe trials," in which the cue for omission of the test stimuli is actually followed by their presentation. The topic is complicated because a number of features of the trial relating to the test stimulus are confounded when the TS is omitted: the presentation of the test stimuli, the opportunity to make a discrimination between them, and the opportunity to obtain reinforcement. There is little doubt that "directed forgetting" diminishes performance, but the reasons are not always clear, and much effort has gone into elucidating them. This topic is treated in some detail by Kendrick and Rilling in Chapter 7, and is also discussed by Wasserman (see Chapter 4). We will describe a less complex feature of working memory trials about which a subject can be informed.

## Anticipated Duration of Memory Intervals

The anticipated duration of the MI is particularly relevant to the study of working memory. This can be studied indirectly on a "steady-state" basis through variations in the distribution of MIs within sessions. A more direct approach is the use of informative stimuli to indicate the duration of the MI on each trial.

Our first encounter with this variable was almost accidental. Honig and Wasserman (1981) compared performance on the DSD and the DCD using, successively, various distributions of the MI: 1, 5, and 10 s; 1, 10, and 20 s; and 1, 20, and 30 s. (Subjects were not informed about the MI value on any trial.) Performance at MIs that were common to two (or all three) distributions varied as a function of the other MI values within the distribution. The 1-s MI was included in each distribution as an "easy" value, but performance even at that value declined when the other MIs were predominantly long (20 and 30 s). It recovered when a more favorable distribution was reinstated. Figure 5.4 provides the opportunity to make comparisons among relevant MI values. A similar effect can be seen in Wasserman's contribution to the same research (see Figure 4.4). When he changed the MI distribution from 0, 5, and 10 s to 5, 10, and 25 s, performance with the 5-s and the 10-s MIs was reduced. Honig and Dodd (1983) obtained a similar effect with their DCD at 6-s and 12-s MIs (see Figure 5.1).

In current research in our laboratory, we are seeking to bring this effect under the control of informative stimuli. Two groups of pigeons were trained in

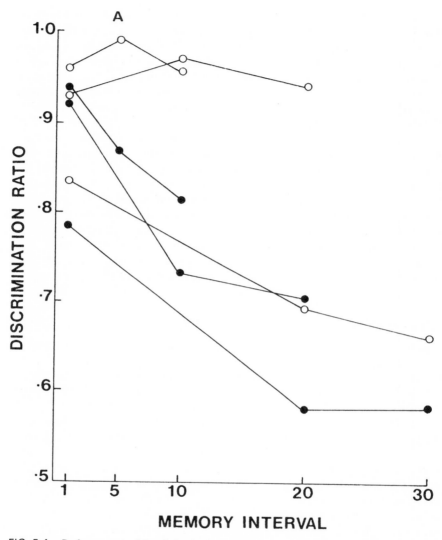

FIG. 5.4. Performance on delayed simple discriminations (open circles) and delayed conditional discriminations (filled circles), with three different distributions of MIs (Honig & Wasserman, 1981).

two DCDs run on separate days in random alternation. In one DCD, the initial stimuli were red and green; in the other, they were white and blue. Line orientations served as test stimuli in both problems. First, the birds were trained to a high criterion of performance (discrimination ratios of .80 or better) in both problems while the MIs were 1 and 5 s. The MIs for one of the two problems were changed to 5 and 10 s. Thus, each pigeon worked with two separate MI distributions that contained a common, intermediate MI value, namely 5 s. The other MI was short in one distribution and long in the other. One group was rewarded only with grain, while the other, the DO group, enjoyed a differential outcome procedure, with water being signaled by one IS in each problem and grain by the other. After 28 sessions with each distribution, the MI values for the DO group were changed to 1 and 9 s as the "short" distribution, and to 9 and 19 s as the "long" distribution. The common MI became 9 s.

Data from the DO group are shown in Figure 5.5. With the first MI distribution, the DRs obtained with the 5-s MI differed at first in accordance with the distribution in which they were contained. As performance on the 10-s MI improved, this effect was reduced. The change in the MI distribution after 7 blocks of sessions had a dramatic effect. Now, the DRs obtained with the 9-s MI differed according to the value of its companion MI. In fact, the values from the 9-s MIs were much closer to their companion values than to each other. The shift to the longer MIs also had a general effect in reducing performance. Even the DRs with the 1-s MI were below the level obtained before the shift.

Data from the other, non-DO group are not provided here, but they are quite similar to those obtained from the DO group following the shift to the longer values. Performance at the intermediate, 5-s MI was affected by the context of

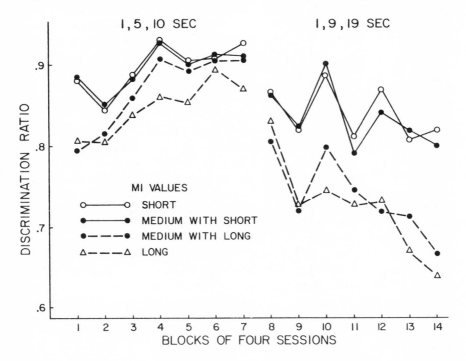

FIG. 5.5.   Two delayed conditional discriminations trained with a common MI; 5 s in the first MI distribution (left), and 9 s in the second MI distribution (right).

the distribution in which that value was contained. Clearly, the anticipations based on general information regarding the distributions of the MIs modulated the memory process to give different levels of performance at a given MI.

## Signaling the Durations of Memory Intervals

The duration of the memory interval (MI) can also be cued on specific trials within a given distribution of short and long MIs. Wasserman, Grosch, and Nevin (1982) describe three such experiments. A DCD was used in the first and second. The compound IS consisted of a color and a line. Colors (green or orange) served as the discriminative stimuli; only "matching" sequences, green-green and orange-orange, led to reinforcement. The informative stimuli were horizontal and vertical lines, which indicated the duration of the upcoming MI, either 1 or 5 s. For one group, line orientations were correlated with MI duration; for another, they were not. The informative cue was compounded with the IS, and then remained on the key during the MI. In each of two replications, the forgetting function was steeper with informative cues; in other words, the effect of the MI was more marked if its duration was signaled in advance.

In a second experiment, subjects were trained only in the correlated condition, using the same general procedure. The experimenters the "miscued" the MI duration on probe trials in the course of two test sessions. The main data from these sessions are shown in Figure 5.6, where "correct" and "incorrect" indicate the validity of the informative cue. Clearly performance was reduced with a short MI when a long MI was cued in advance, and, conversely, when the long MI was miscued as a short one, discrimination improved. In fact, the memory function has a slight positive slope.

We carried out similar research independently (Dodd & Honig, 1981), using the DSD rather than the DCD, and MIs quite a bit longer than those used by Wasserman et al (1982). Horizontal and vertical lines served as the initial stimuli, with food available for responding to a white test stimulus (TS) after one line orientation but not after the other. Colors as informative cues were combined with the IS. Two colors signaled the upcoming MI as short or long; a third color provided no differential information about the MI. These informative cues ended with the IS. Training proceeded at first only with a short MI of 5 s; then a second, long MI was added, and this was adjusted to 20 or 30 s for different birds to assure a performance better than chance but poorer than with the short MI. In order to enhance the difference between MI durations, the short MI was then changed to 0 s. When the pigeons were informed of the MI duration, discrimination was somewhat better following a 0-s MI and somewhat worse following the long MI than it was in the unsignaled condition (see the left panel of Figure 5.7). While the slope difference is small, it is reliable.

Following this training procedure, probe trials were administered, in which a cue for the short MI preceded the long MI, and vice versa. Compared to

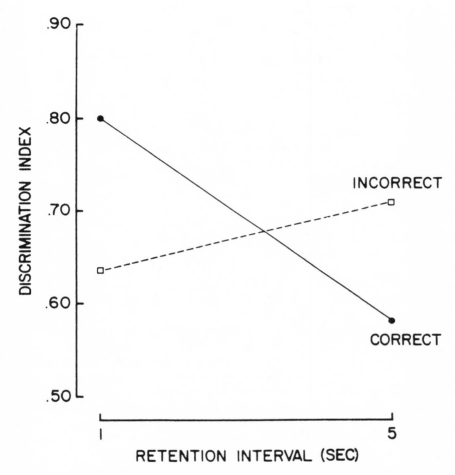

FIG. 5.6.   Forgetting functions from a delayed conditional discrimination in which informative cues signaled the duration of the MI (Wasserman et al., 1982).

correctly cued trials, significantly poorer discrimination ratios were obtained on short-MI trials that began with the long-MI cue, and somewhat better discrimination ratios were obtained on long-MI trials that began with the short-MI cue. The latter difference was small and not significant (see the second panel of Figure 5.7).

   In a systematic replication of this study, the MI was signaled on all trials by the color on the key. The short MI was 1 s, and the long MI was 20 or 30 s. The main objective was the replication of the probe test, but all informative cues were correlated with MI duration. The average discrimination ratios obtained from the last five sessions of training are shown in the third panel of Figure 5.7, and the ratios from the probe session are at the right. There was a significant

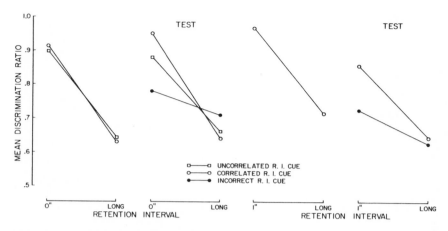

FIG. 5.7.   The effects of informative and uninformative cues about MI duration in a delayed simple discrimination (Dodd & Honig, 1981).

reduction of discrimination on short-MI probe trials following the long-MI cue, in comparison with correctly cued trials. On long-MI trials there was no effect of the color cue.

These experiments, by Wasserman et al. (1982) and by Dodd and Honig (1981), are similar in purpose and design and provide comparable data. The main difference is our failure to obtain enhanced performance when the long MI was cued as a short one. With a DSD, we could use a much longer "long" MI than did Wasserman et al. The length of this MI may have overridden any effect of the cue indicating that it would be short. Furthermore, our informative cues ended with the IS, while Wasserman et al, extended theirs through the MI. Our subjects had to remember the informative cues, while the others did not.

## Unsignaled, Unanticipated Memory Intervals

Pigeons seem to anticipate a general distribution of MI values in the absence of specific informative cues, so they might also come to anticipate specific MI durations within such a distribution. If so, performance following a new, "unexpected" MI could be reduced relative to that obtained after the "standard" MIs. This would be particularly convincing if the unanticipated MI is shorter than the standard values. Dodd, in unpublished research, trained pigeons extensively with a DSD that always involved 5 s as the short MI and 30 or 40 s as the long one. Training continued until performance was quite good at the 5 s MI and reasonably good at the long MI. (The specific value of the long MI was chosen to meet this objective.) After 50 sessions with the final MI values, two probe sessions were conducted in which three other MI durations were presented on randomly selected trials. These were 1, 15, and 50 s; thus they were short, intermediate, and long, relative to the MIs that the birds had experienced.

Performance with the 1-s MI probe trials was poorer than it was with the 5-s MI in both probe sessions. Performance on the 15, 30/40 and 50-s MIs generally declined as the values increased. The decline at the 1-s MI was of particular interest. It may be argued that this decline was only due to some behavior such as turning away briefly from the key after the IS. Nevertheless, such mediating behavior may reflect the process of interest to us, namely the anticipation that no test stimulus will be presented until a certain time has elapsed.

## Discussion

These studies indicate that pigeons are affected by the anticipation of MI durations, whether these are signaled on a trial-by-trial basis, signaled as a distribution of MI durations, or just put in effect as a set of specific intervals during many training sessions. Anticipation of shorter MIs enhances working memory, while anticipation of longer ones reduces it. This effect is found with various training procedures: the DCD, the DSD, and DMTS. The prospective nature of such a process is supported most convincingly if the informative cues are not continued through the MI.

The mechanism of action of informative cues has not been analyzed, nor do we have information on the remembering of informative stimuli. When such cues are presented together with the initial stimuli, or in advance of them, they may affect memory of the initial stimuli only indirectly, by controlling the degree to which the subject attends to the IS, or "encodes" it. The pigeon may pay more attention to the IS if it can anticipate a short MI or a large reward. The effect of informative stimuli on memory would best be carried out if such cues are presented after the initial stimuli.

Little is known about memory for informative cues. To study this, it would probably be more useful to provide information about a variable other than MI duration. Since the criterion performance is so strongly determined by the MI, studies of the modulation of performance on this dimension with stimuli that may themselves be forgotten would be complicated indeed. On the other hand, a signal for the size of the reward may also affect performance, and the effects of this signal may be subject to the time between its occurrence and the presentation of test stimuli. In principle, informative cues can be presented at any time during the MI.

Our account of informative cues points out an important feature of initial stimuli: They may also have an informative function, simply because they precede important features of the trial. Our work on MI distributions, and Dodd's unpublished work on unanticipated MIs, support the informative function of initial stimuli. In differential outcome studies, the IS is sufficient to cue the correct response, but it also acts as an informative cue because it indicates features of the trial outcome. One of the advantages of separating informative and initial stimuli is that the former can take over the informative function of the latter. This would allow us to study prospective and retrospective remembering without the influence of the anticipation of trial characteristics.

## CONCLUSION

We have reviewed a modest set of anticipations that animals—pigeons in particular—can develop in the course of working memory experiments. These anticipations can affect responding after the memory interval (MI), and in general we detect them by comparing the accuracy of responding in the different conditions that are run in several experimental paradigms. No doubt there could be more sensitive indicators, especially in the actual form of the response. Unfortunately, there has been little direct observation of anticipatory behavior, such as orientation toward a test stimulus (or its anticipated location), "food pecking" versus "water pecking," and so forth. Such observations might give us a clue regarding the nature or "content" of the anticipations. At this time, the content has to be inferred from the interaction between accuracy of responding to test stimuli and the information provided in the training paradigm. Thus, we suggest that pigeons generate *response intentions* if information regarding the correct response is available from the beginning of the trial, but the trial outcomes for correct responding are all the same. We infer *outcome expectancies* if rewards are differentially correlated with initial stimuli. Pigeons generate *anticipations of trial characteristics* when aspects of a trial are varied independently of the stimulus/response contingencies that define correct responding. In all cases, these inferences are based upon comparing behavior in conditions where such anticipations can or cannot be generated.

We need hardly repeat our view that anticipations are prospective memory processes. As such, intentions and differential outcome expectancies improve performance after a delay, while anticipations of trial characteristics have mixed effects. Presumably, anticipations are also subject to forgetting. The data that we have reviewed indicate that intentions and outcome expectancies are forgotten less quickly than the initial stimuli which generate them, when the latter serve as cues in a conditional discrimination. Little is known about the forgetting of trial characteristics when they are cued by informative stimuli. We also need to learn about the process by which anticipations of trial characteristics modulate performance, whether they affect the processing or "encoding" of initial stimuli, or the rate of forgetting.

If all anticipations and intentions are prospective, we may speculate upon their modes of action, and upon the differences among these modes of action. Differential outcome expectancies apparently act as surrogate cues for initial stimuli. We have already documented this conclusion in our description of transfer studies based on such expectancies. In the case of DMTS they cue the choice of one TS over another. But in the DCD—the paradigm with which we have worked—they cue differential rates of responding to separate test stimuli. As pigeons are biased to peck at a single stimulus for a possible reward, low DR values are due to a failure to withhold responding when the TS is negative. This leads to an interesting conclusion: A specific outcome expectancy (food versus water, for example) mediates suppression of responding to obtain that outcome when the TS is presented in a negative trial, and does so more effectively than does the memory of the IS. This is rather remarkable in view of the difficulty of obtaining suppression of key-pecking in "omission training," when a cue that precedes reinforcement instructs the pigeon not to peck before it is presented.

We have suggested that superior performance in the DSD (compared to the DCD) may also be mediated by an expectancy generated by the IS. Anticipation of reward or its omission can be generated by the initial stimuli, and this can result in different response intentions to the TS. Outcome expectancies would then be mediating response intentions.

In contrast, different response intentions from a simple discrimination do not appear to mediate responding within a different discrimination. In Whalen's (1979) study, the same initial stimuli cued differential response rates to a single TS to gain reward, and they cued a choice between two test stimuli in DMTS. The response intention, which we assume to be mediating the DSD, could in principle have cued the choice in DMTS, but it did not. There are several confounded differences between Whalen's experiments and the work on outcome expectancies. Among others, the assessment of the mediating function was quite different. Whalen was looking for improved memory in DMTS without special memory training to support mediation based on response intentions. He did not use the sort of transfer design used for studying outcome expectancies.

It would be of interest to develop a transfer design parallel to those used in differential outcome studies, but using different response patterns instead. The same response intentions would be trained with independent sets of initial stimuli, each of them cueing choices within separate sets of test stimuli, and different response patterns to the correct stimuli. The test stimuli could then be interchanged in the manner of Edwards et al. (1982). If transfer based on response intentions is not obtained, this would suggest that outcome expectancies do not depend upon the anticipated responses to the trial outcomes: eating versus drinking, eating corn versus eating wheat, eating versus not eating, and so forth. The mediation would depend on the expectancies themselves, some representation of the reward as such.

Anticipations of trial characteristics can modulate responding after the MI, as we have seen; in our standard example, the instruction for a short MI enhances performance, while the instruction for a long MI reduces it. The question arises naturally whether such anticipations can mediate differential responding so that performance could be improved following both of the stimuli typically used in an "informative" design. Normally, the informative cues are orthogonally related to the initial stimuli in order to avoid confounding these sets of cues. If one were to correlate the informative cues with the test stimuli in such a way that they, or expectancies generated by them, could mediate responding, then they would be redundant with the initial stimuli. But then they would be superfluous. The procedure would be in effect a differential outcomes procedure, since informative cues normally provide information relevant to some aspect of the reinforcer or trial outcome. Signaling differences in the MI, for example, inevitably entails the signaling of differences in the time between the IS and the reinforcer. Complex and ingenious designs would be required to eliminate this confound, if it can be done at all.

Since it seems to be difficult to use informative stimuli as cues to facilitate correct responding, it may be useful to determine whether they can interact with initial stimuli to obtain the opposite effect, a reduction in stimulus control. The information provided by informative cues may overshadow the control by the initial stimuli. One could, for example, use differential outcomes, but cue them with informative rather than with initial stimuli. Colors could serve as initial and as test stimuli, and line orientations could indicate whether the trial outcome will be food or water. In a control procedure, line orientations would be randomly associated with the trial outcome. If performance is poorer with the informative procedure, this suggests that the anticipation of the trial outcome can overshadow the memory for colors by the initial stimuli. On the other hand, this anticipation may make the outcomes more discriminable and thus assist performance, even if the anticipation is not relevant to the correct response and cannot serve as a cue.

We have not tried to resolve the question of the stimulus "content" of the intentions and anticipations that we have discussed. This question probably cannot be answered with any confidence. What is represented in anticipations, stimuli or responses? The content of a response intention is presumably a pattern of responding, but different response patterns also produce different stimuli. If subjects are sensitive to the external stimuli produced by their own behaior, rather than to the proprioceptive feedback, then even a response intention may incorporate stimulus content. Differential outcomes are of course stimuli, but they generate different response patterns, which can be anticipated and could contribute to the expectancies. The anticipation of trial characteristics is not well understood, but much the same can be said of them. The anticipation of a set of cues or of a particular MI may be coded as a stimulus, but since "mediating" behaviors can never be eliminated, the content of the anticipation cannot be specified with any certainty.

The stimulus content of anticipations could be approached more directly by extending the domain of relevant research. We have already mentioned the directed-forgetting procedures, which involve the omission or replacement of test stimuli. Informative cues can also be associated with alternate sets of test stimuli. If these are "miscued," performance may be reduced. The literature on this topic is sparse (Riley & Roitblat, 1978; Roitblat, 1981; Stonebraker, 1981; see also Chapter 4). The anticipation of specific stimuli remains to be explored as a process that may affect working memory in animals.

## ACKNOWLEDGEMENTS

Original research reported in this chapter was supported by grant no. AO102 from the National Research Council of Canada. We are indebted to Dawn Davis, William Matheson, Shirlene Sampson, and Barbara Vavasour for experimental assistance, and to Marcia Spetch for comments on an earlier version of this chapter.

## REFERENCES

DeLong, R. E., & Wasserman, E. A. Effects of differential reinforcement expectancies on successive matching-to-sample performance in the pigeon. *Journal of Experimental Psychology: Animal Behavior Processes*, 1981,7, 394–412.

Dodd, P. W. D., & Honig, W. K. Effects of signalling the duration of a retention interval in a delayed discrimination. Paper presented to the Canadian Psychological Association, June 1981.

Edwards, C. A., Jagielo, J. A., Zentall, T. R., & Hogan, D. E. Acquired equivalence and distinctiveness in delayed matching to sample by pigeons: Mediation by reinforcer-specific expectancies. *Journal of Experimental Psychology: Animal Behavior Processes*, 1982, *8*, 244–259.

Honig, W. K. Studies of working memory in the pigeon. In S. H. Hulse, H. Fowler, & W. K. Honig (Eds.), *Cognitive processes in animal behavior*. Hillsdale, NJ: Lawrence Erlbaum Associates, 1978.

Honig, W. K., & Dodd, P. W. D. Delayed discriminations in the pigeon; The role of within-trial location of conditional cues. *Animal Learning and Behavior*, 1983, *11*, 1–9.

Honig, W. K., Matheson, W., & Dodd, P. W. D. Outcome expectancies as mediators for discriminative responding. *Canadian Journal of Psychology*, 1984, *38*, 196–217.

Honig, W. K., & Thompson, R. K. R. Retrospective and prospective processing in animal working memory. In G. H. Bower (Ed.), *The psychology of learning and motivation, Vol. 16*. New York: Academic Press, 1982.

Honig, W. K., and Wasserman, E. A. Performance of pigeons on delayed simple and conditional discriminations under equivalent training procedures. *Learning and Motivation*, 1981, *12*, 149–170.

Peterson, G. B. The differential outcomes procedure: A paradigm for studying how expectancies guide behavior. In H. L. Roitblat, T. G. Bever, & H. S. Terrace (Eds.) *Animal cognition*. Hillsdale, NJ: Lawrence Erlbaum Associates, 1984.

Peterson, G. B., & Trapold, M. A. Effects of altering outcome expectancies on pigeons' delayed conditional discrimination performance. *Learning and Motivation,* 1980, *11,* 267–288.

Peterson, G. B., Wheeler, R. L., & Trapold, M. A. Enhancement of pigeon's conditional discrimination performance by expectancies of reinforcement and nonreinforcement. *Animal Learning and Behavior,* 1980, *8,* 22–30.

Riley, D. A., & Roitblat, H. L. Selective attention and related cognitive processes in pigeons. In S. H. Hulse, H. Fowler, & W. K. Honig (Eds.). *Cognitive processes in animal behavior.* Hillsdale, NJ: Lawrence Erlbaum Associates, 1978.

Roitblat, H. L. The meaning of representation in animal memory. *The Behavioral and Brain Sciences,* 1982, *5,* 353–372.

Smith, L. Delayed discrimination and delayed matching in pigeons. *Journal of the Experimental Analysis of Behavior,* 1967, *10,* 529–533.

Stonebraker, T. B. *Retrospective and prospective processes in delayed matching to sample.* Unpublished doctoral dissertation, Michigan State University, 1981.

Wasserman, E. A., Grosch, J., & Nevin, J. A. Effects of signaled retention intervals on pigeon short-term memory. *Animal Learning and Behavior,* 1982, *10,* 330–338.

Whalen, T. E. *Support for a dual encoding model of STM in the pigeon.* Unpublished doctoral dissertation, Dalhousie University, 1979.

# 6 Proactive Interference in Animal Memory

Anthony A. Wright
*University of Texas Health Science Center, Graduate School of Biomedical Sciences*

Peter J. Urcuioli
*Purdue University*

Stephen F. Sands
*University of Texas at El Paso*

In this chapter, a new perspective on the role of proactive interference (PI) in animal memory is presented. We will argue that repetition of the events that animals must remember during an experimental session generates a level of PI much greater and much more detrimental to performance than has been previously realized. Consequently, we believe that short-term memory capabilities of animals are often grossly underestimated, and that more accurate assessments of these capabilities require elimination of item repetition (and accompanying PI) within a session. Our discussion of PI will be limited to two particular memory paradigms, delayed-matching-to-sample (DMTS) and serial-probe recognition (SPR), but our message should be applicable to other situations as well.

Interference from prior events on memory for currents ones (viz. PI) has typically been viewed as an effect confined primarily to the immediately adjacent trials of a memory task (e.g., Hogan, Edwards, & Zentall, 1981; Moise, 1976) or, sometimes, within individual memory trials themselves (e.g., Grant & Roberts, 1973). How PI adversely affects overall memory performance is implicitly tied to this view, perhaps because it is easy to demonstrate within-trial or across-trial PI. We will argue, however, that a larger background of PI is present throughout the entire experimental session, and that the effects seen within trials or across successive trials represents only a relatively small increment in PI above this background.

Most experimenters repeatedly present just a few memory items within an individual session, and across daily sessions. We think that this continual repetition suppresses performance throughout an *entire* session, not just across successive trials. We call this general suppression the *repeated-item PI*

101

*effect,* to distinguish it from interference measured solely within individual trials or across immediately adjacent trials. Our evidence for repeated-item PI is based on the rather poor short-term memory performance exhibited by subjects performing DMTS and SPR with small numbers of repeated items (Gaffan, 1977; MacPhail, 1980), and on data showing dramatic improvement in memory performance when such repetition is minimized or eliminated (Sands & Wright, 1980a, 1980b; Overman & Doty, 1980).

We have organized the chapter into three sections. In the first section, we review how PI has typically been studied in DMTS, the paradigm which has provided the majority of data on interference in animal memory. We will question whether the estimates of PI provided by these studies truly reflect how much prior events interfere with memory for current ones. In the second section, we extend our analysis to the SPR task, where the detrimental effects of repeated-item PI can make the study of multiple-item memory in nonhuman animals virtually impossible. In the last section, we summarize, integrate, and theorize about the mechanism of the repeated-item PI effect.

## PROACTIVE INTERFERENCE IN DELAYED MATCHING-TO-SAMPLE

Subjects in DMTS typically choose between two comparison stimuli on each trial, one of which has previously appeared as the sample (e.g., Blough, 1959). Events occurring prior to the sample can potentially interfere with remembering during the delay interval which follows sample presentation and precedes the choice (or retention) test. According to the "traditional" view of PI in DMTS (see, for example, Roberts & Grant, 1976), interference arises when the retention test involves a choice between the sample for the current trial and the one from the immediately preceding trial. If the prior sample continues to be remembered, the choice alternatives are "conflicting" and, not surprisingly, errors tend to be relatively more frequent than when the retention test does not involve such a conflict. Implicit in this analysis is that PI is primarily confined to adjacent DMTS trials.

Most studies of DMTS, however, use only a few (often only two) samples within each DMTS session, so one can reasonably argue that virtually every retention test involves a conflict between memory items (cf. D'Amato, 1973). If so, then performance on any particular trial may suffer not only from events on an immediately preceding trial but also from events on other previous trials as well. Interference would be most acute when only two samples are used because, in this limiting case, *every* retention test (with perhaps the exception of the first one or two) involves conflicting choices.

In the next section, we review the assessment of PI in DMTS from the standpoint that it is a relatively "confined" or localized effect. Later, we examine some data which suggest that a much more pervasive PI effect is produced by the repeated presentation of a small number of stimuli within a session.

## Traditional Assessments of PI in DMTS

The influence of PI on DMTS performance has usually been assessed by examining the effects of previous trials on current trial accuracy, or by observing how performance is affected by the explicit presentation of stimuli immediately prior to the sample. The former method of assessment, called the *intertrial PI procedure,* does not involve any experimental manipulation. Accuracy on Trial $N$ is simply computed as a function of whether the sample (or comparison chosen) on the preceding trial (Trial $N_1$) differed from the current sample (or current correct comparison). PI is implicated when accuracy on Trial $N$ is lower and when the previous trial employs the alternative sample/ correct comparison. The latter method of assessment, called the *intratrial PI procedure,* involves an effect whereby a stimulus presented immediately prior to the sample (a presample stimulus) reduces matching accuracy relative to that seen on control trials in which this additional stimulus is absent.

*Intratrial PI.*    Studies of intratrial PI have consistently shown that presentation of a presample stimulus produces an interference effect which can be reproduced with different species, stimulus sets, and training procedures. Grant and Roberts (1973), for example, demonstrated that choice accuracy of pigeons performing DMTS with red, green, blue and yellow hues was reduced at all retention intervals when the alternative sample (viz., the incorrect choice on the retention test) immediately preceded the sample to be matched. Grant (1982) and Zentall and Hogan (1974, 1977) have replicated these effects in pigeons; Jarvik, Goldfarb, and Carley (1969, Experiment 3) have demonstrated them in monkeys. Medin (1980, Experiment 1) reported that presample presentations of the incorrect choice stimulus also disrupted memory performance by monkeys trained on a DMTS task involving novel three-dimensional objects on each trial (viz. trial-unique stimuli).

Intratrial PI is apparently specific to the presentation of the alternative sample/incorrect choice stimulus. For example, Zentall and Hogan (1974, 1977) and Medin (1980) showed that an irrelevant presample stimulus (one which does not appear on the retention test) does not interfere with performance. This finding is consistent with the notion that PI results from *conflicting* memory items: namely, items which appear together on the retention test.

The size of the intratrial PI effect, however, is often quite small (at least with pigeons as subjects). Accuracy is usually reduced by only 5-8%, although such a small effect might in some instances be attributed to a floor effect: Accuracy on control trials is often so low that little room is left to detect any sizeable disruption. For instance, the pigeons in the Grant and Roberts (1973) experiment matched correctly on only 60-70% of control trials at delays longer than 0 s. (Note that chance performance is 50% correct.) Similarly, control-trial accuracy for the pigeons in the Zentall and Hogan (1977) experiment was only 60-65%. This poor level of baseline accuracy makes it hardly surprising that the intratrial PI performance decrement was quite small.

We think that the relatively poor performances on no-interference control trials is really the more significant result in these studies. The remarkably poor retention by subjects even with delays of only a few seconds implies to us a much more substantial and pervasive PI effect, one which results from the repeated presentation of a few memory items within a session. Consistent with this view are data showing that monkeys can achieve very high levels of DMTS accuracy (viz. 90% correct or better) at relatively long delays when the stimuli they must remember are unique on each trial (Medin, 1980; Overman & Doty, 1980). Interestingly, intratrial PI under these conditions appears to be a much larger effect.

*Intertrial PI.*    Intertrial PI is assessed by comparing matching accuracy on target trials as a function of what has occurred on preceding trials. If there is a change from one sample to another (and thus a reversal in the correct and incorrect choices) across two successive trials, the second (target) trial is part of a negative trial-transition. When the sample remains the same (as do the correct and incorrect choices), it is part of a positive trial-transition. The former is assumed to produce PI by creating a conflict in choice on the target trial. Data consistent with this view can be seen in Figure 6.1, which shows

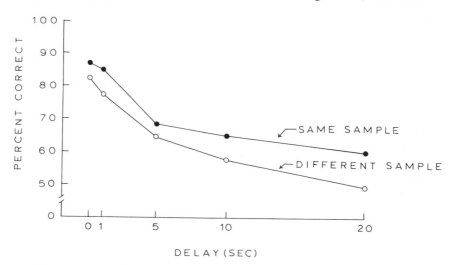

FIG. 6.1.   Accuracy of delayed matching by monkeys on trials following those with the same sample or with a different sample (After Moise, 1976).

accuracy of monkeys on the target trials of positive and negative transitions reported by Moise (1976). Performance was consistently less accurate on negative transitions than on positive transitions across the five delay (retention) intervals in Moise's experiment. Similar effects have been obtained from dolphins performing auditory DMTS (Herman, 1975; Thompson & Herman, 1981) and from pigeons matching hue stimuli (Grant, 1975). Furthermore, monkeys have been shown to be less accurate on DMTS as the number of negative transitions within a session increases (Worsham, 1975).

The difference in performance between positive and negative trial-transitions is apparently a true interference effect and not a facilitation effect due to repeating the same sample/correct choice on positive transitions. For example, Hogan, Edwards, and Zentall (1981) and Grant (1975, Experiment 2) found that DMTS performance by pigeons was worse on negative transitions than on neutral transitions in which the choice alternatives (or at least the correct choice alternative) from the preceding trial was absent on the target trial. Positive transitions did not produce a corresponding facilitation relative to the neutral control. These data coincide very nicely with results in the human literature (Bennett, 1975; Wickens, Born & Allen, 1963) indicating that inferior performance on negative transitions is a true PI effect.

The outcome of a prior trial (i.e., reward or nonreward) might also be expected to influence the amount of intertrial PI, and it does. Moise (1976), for example, found that monkeys performed more poorly on target trials of negative transitions if they selected the correct (rewarded) alternative on Trial $(N-1)$ than if they selected the incorrect (nonrewarded) alternative. This effect can be understood by noting that in the rewarded condition, the potentially interfering stimulus is seen twice (as the sample and choice stimulus) whereas, in the nonrewarded condition, it may be seen only once. Alternatively, reward for a previous choice may create a bias for the same choice on the next trial. Consistent with this latter interpretation was the additional finding by Moise (1976) that monkeys were more accurate on target trials of positive transitions when they selected the correct (rewarded) alternative on Trial $(N-1)$ than when they selected the incorrect (nonrewarded) alternative.

Roberts (1980) and Roitblat and Scopatz (1983) have suggested yet another interpretation of this pattern of results. They argue that data such as these are consistent with the view that PI in DMTS arises from the previously selected choice stimulus (rather than from the previously seen sample), and that the effect is independent of whether the prior choice is rewarded or not. Their point is illustrated in Table 6.1 which shows sequential-effects data from Roberts (Experiment 1) for pigeon DMTS and from Moise (1976) for monkey DMTS. Accuracy on Trial $N$ is broken down as a function of the sample on Trial $(N-1)$ (the usual intertrial analysis), the Trial $(N-1)$ outcome, and the choice stimulus selected on Trial $(N-1)$.

Subjects that made a correct (rewarded) choice on Trial $(N-I)$ were less accurate on Trial $N$ if the sample was different from the previous trial than if it

TABLE 6.1
Trial-transition Effects in Identity DMTS as a Function of Prior
Sample, Prior Reward Outcome, and Prior Choice

| Trial (N–1) | | % Correct on Trial N | | Trial (N–1) Choice vis-a-vis Trial N Sample |
|---|---|---|---|---|
| Sample | Outcome | Roberts (1980) | Moise (1976) | |
| Same | R | 73.6 | 75.3 | Same |
| Same | NR | 68.8 | 64.5 | Diff |
| Diff | R | 65.1 | 62.3 | Diff |
| Diff | NR | 73.4 | 71.4 | Same |

Note: Diff = different, R = reward, NR = nonreward

was the same (compare rows 1 and 3). This is the typical negative trial-transition effect. Just the reverse is true, however. Subjects that made an incorrect (nonrewarded) choice on Trial (N–1) were more accurate on Trial N if the sample was *different* from that on the previous trial (compare rows 2 and 4). Apparently, subjects momentarily perseverate in their choices across trials, whether or not these choices have been previously rewarded.

This can be appreciated by referring to the righthand column of Table 6.1, which reclassifies the Trial N sample as "same" or "different" relative to the choice stimulus selected on the prior retention test. The reclassification shows that accuracy is lowest when subjects have just selected a choice stimulus which is different from the sample for the current trial (cf. middle two rows). Selection of the alternative choice stimulus on the prior retention test is indicated when either: (1) two successive samples are different and the Trial (N–1) outcome is reward (row 3); or (2) the samples are the *same* on two successive trials and the Trial (N–1) outcome is nonreward (row 2). In the latter situation, one might expect that nonreward on Trial (N–1) would bias subjects towards choosing the alternative stimulus on Trial N (which, on positive trial-transitions, would be correct), but apparently this is not the case. Subjects seem to perseverate in their choices, independently of prior reward or nonreward, and this momentary perseveration appears to be the source of intertrial PI.

Why should a prior choice affect memory for a current sample? One explanation is that subjects code the sample in terms of the correct choice stimulus at the beginning of a trial and remember this prospective event during the retention interval (cf. Honig & Thompson, 1982; Roitblat, 1980). The choice stimulus selected on the previous trial can then potentially interfere with memory for the current correct choice. Regardless of the actual mechanism underlying intertrial PI, however, it is perhaps more important to note that the effect is usually quite small (e.g., on the order of a 5–8% increase in accuracy). If this estimate reflects the overall effect of PI in DMTS, then one must conclude that the relatively poor performances by subjects in this task indicate rather limited short-term memory capacities: perhaps only 6–10 s in highly trained pigeons and 1–2 min. in highly trained monkeys and dolphins. By these delays, accuracy has usually fallen to chance. We prefer to think, however, that animals are capable of remembering much more accurately over much longer delays and that a greater-than-estimated PI effect is responsible for poor performances.

*Intertrial Interval Effects.*    A corollary of the intertrial analysis is that amount of PI should vary with the intertrial interval (ITI). Long ITIs should reduce PI because competing memories from a prior trial are more likely to be forgotten (cf. Loess & Waugh, 1967). Conversely, short ITIs should increase PI because relatively less forgetting will occur. These predictions have generally been supported in the literature. Hogan et al. (1981), for example, found that pigeons performing DMTS with 1-s delays were less accurate on negative than on neutral transitions when ITIs were either 2 or 5 s but not when they were 10 s. Grant (1975, Experiment 4) reported significant intertrial PI with pigeons when the ITI was 2 s but not when it was either 20 or 40 s. Furthermore, overall accuracy in DMTS has been found to be directly proportional to the length of the ITI in pigeons (Maki, Moe & Bierley, 1977; Roberts, 1980), monkeys (Jarrad & Moise, 1971; Mason & Wilson, 1974), and dolphins (Herman, 1975, Experiment 1).

According to some views (e.g., D'Amato, 1973), performances should be influenced by the *ratio* of the ITI to the average retention interval (rather than to the ITI per se), because subjects must discriminate periods during which they should be remembering from those during which remembering is not required. As the ITI increases with a constant delay, this temporal discrimination should become easier. As it decreases, the discrimination should become more difficult and thus produce more intertrial interactions (viz. PI). Of course, the influence of temporal cues is probably most pronounced when the stimulus conditions of the retention interval are identical or similar to those of the ITI (e.g., Roberts & Kraemer, 1982).

Recently, Roberts (1980) has questioned whether the ITI affects DMTS performance via PI or by simply decreasing attention to the samples. He noted that shorter ITIs not only produced lower accuracy on the typical, two-sample (heterogeneous) DMTS task, but also reduced accuracy when the sample and correct choice were the same on every trial (homogeneous DMTS). Supposedly, interference should not be present in the homogeneous DMTS where short ITIs might be expected to facilitate performance (cf. Roberts & Kraemer, 1982; Williams, 1971). These data are intriguing, but we wonder whether the results from the homogeneous task may have been influenced by the birds' extensive prior experience on heterogeneous DMTS. Also, interference from previous trials *was* evident in homogeneous DMTS: The source was the side key most recently pecked. Finally, Roberts's suggestion that the ITI may simply affect processing of the sample stimulus seems to be at odds with human data (Gardiner, Craik & Birtwistle, 1972; Watkins & Watkins, 1975) showing that PI is primarily a retrieval deficit not a storage deficit.

### Effects of Sample Set Size

The generally poor performance of subjects on DMTS with delays of only a few seconds suggests that the repetition of stimuli produces a much larger, more pervasive PI effect than previously thought. This larger effect might develop if PI accumulates across successive memory trials (Keppel & Underwood, 1962; Petrusic & Dillon, 1972). In other words, performance on, say, Trial 30 may be depressed as a result of experience with all prior 29 trials rather than just by Trial 29 alone, although it is only the latter which has been typically used to estimate PI. Alternatively, since subjects are tested repeatedly over days with the same stimuli, a substantial amount of PI may be generated within a few trials if the alternatives on each retention test act as retrieval cues for conflicting memories (Gardiner et al., 1972).

In any event, overall accuracy in DMTS should improve substantially with larger numbers of samples and comparisons, because the frequency of conflicting choices on the retention tests will be reduced (cf. Mason & Wilson, 1974). In the limit, accuracy should be highest using samples and comparisons which appear only once within a session. In other words, PI should be virtually absent with trial-unique stimuli because, theoretically, there will be no conflicting choices on any retention test. Of course, this will depend to some degree upon how similar the members of the memory set are to each other: Conflicts might still arise if an incorrect choice resembles a prior sample.

Studies of the effects of sample-set size on DMTS show that increasing the number of sample stimuli does dramatically improve memory performance. Mishkin and Delacour (1975, Experiment 1), for example, trained monkeys on a 10-s DMTS task with either a single pair of three-dimensional junk objects or with trial-unique objects. They found that the trial-unique group acquired the task to criterion (90% accuracy or better for two consecutive 20-trial sessions) twice as rapidly as the single-pair group. Furthermore, when the trial-unique group was shifted to the single-pair condition after reaching criterion, accuracy was immediately reduced by 25% and remained at this lower level for 20 sessions. The inability to maintain accurate performance with a pair of familiar stimuli when the pair was repeated within a session is strong evidence for a repeated-item PI effect we beieve is common to many DMTS tasks.

Overman and Doty (1980) compared peformances of monkeys trained sequentially on two DMTS tasks: one involving trial-unique stimuli consisting of a set of 35 mm color slides which were rear-projected onto three horizontally aligned screens, and the other involving a single pair of stimuli drawn from this set and repeatedly presented within a session. Figure 6.2 shows that accuracy was considerably higher with trial-unique than with repetitive stimuli at all

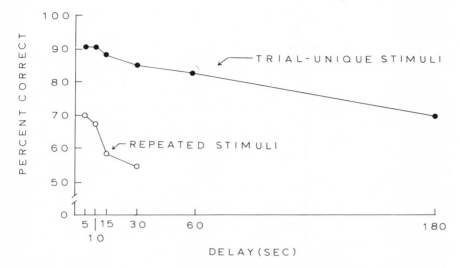

FIG. 6.2.  Accuracy of delayed matching by monkeys with either a single pair of stimuli comprising the sample set for each session (repeated stimuli) or with sample stimuli appearing only once within each session (trial-unique stimuli) (After Overman and Doty, 1980).

delays. Even more impressive was the finding that monkeys were able to remember the trial-unique stimuli at better than chance levels of accuracy with delays as long as 24 hours (not shown in the figure), while they were no better than chance at delays of only 30 s with repeated items. Repeated-item PI seems to be a good candidate for explaining these huge peformance differences.

Worsham (1975) and Mason and Wilson (1974) have also reported better accuracy with increasing sample-set sizes, although their effects were not as large as those reported by Overman and Doty (1980). The smaller effects may have been due to their smaller pools of items with unidimensional characteristics, which would tend to produce across-item generalization. Nonetheless, the direction of their effects are consistent with those cited above.

To date, only monkeys have been used in this line of research. There is no a priori reason, however, to suppose that the effects would be any different for any other species. Research reproducing these results in other species is certainly needed. In any event, the data currently available strongly suggest that repetition of sample stimuli in DMTS produces relatively poor memory performance, not because subjects forget the samples over the retention intervals (cf. Wilkie & Spetch, 1981), but rather because, in some sense, they remember the samples too well and are unable to distinguish *which* memory item was presented at the start of each trial (D'Amato, 1973). This powerful repeated-item PI effect can be minimized or eliminated, and performance improved, by using stimuli which are not repeated within a session.

## PROACTIVE INTERFERENCE IN LIST MEMORY TASKS

A procedure recently developed to study multiple-item memory in monkeys is the serial-probe-recognition (SPR) task (Sands & Wright, 1980a, 1980b, 1982). In this task, the monkey initiates each memory trial by pressing down on a three-position response lever. The items to be remembered are then sequentially presented on the top of two vertically aligned viewing screens. Immediately following the last item in the list, a probe or test item is presented on the bottom screen. The monkey's task is to classify the probe as either identical to one of the list items or different from all of them. It does so by moving the response lever to the right ("same") or the left ("different") respectively (cf. Sands & Wright, 1980a for more experimental details). Proactive interference can occur when a probe item not appearing in the current list (viz., a "different" probe) appears as a member of some previous list. In this situation, the monkey may be confused and respond "same" to the different probe of the current list, demonstrating across-list confusion (viz. PI).

### Repeated-item PI in Multiple-item Memory

The effect of repeating items in list-memory tasks can be demonstrated by comparing performances of subjects under two training conditions: one in which the stimuli comprising the lists are regularly repeated within a session, and another in which the stimuli are trial-unique. In our laboratory, list lengths of three items were used to make this comparison because lists of this particular size had been used in several previous studies of multiple-item memory (e.g., Devine & Jones, 1975; Eddy, 1973; Gaffan, 1977). All of these previous studies employed a small pool of items from which to create three-item lists; consequently, the memory items in these studies were regulary repeated within each session. This should, of course, create a condition of high PI. In our experiments the high-PI condition was reproduced by constructing three-item lists from an item pool containing only six items. The comparison condition which we ran was one in which the three-item lists were constructed from a pool containing 211 different items, and list items were not repeated within a session (low-PI condition). Monkeys were tested on eight 140-trial sessions with four sessions devoted to each of these two conditions.

Figure 6.3 shows that accuracy on the SPR task for both a rhesus monkey and a human tested under similar conditions was substantially lower under the high-PI than under the low-PI condition. This difference in performance is even more impressive considering that the six items in the high-PI condition were the best discriminated items from the 211-item pool (Sands & Wright, 1980a). We have recently replicated these results with four other rhesus monkeys; their averaged data appear in Figure 6.4.

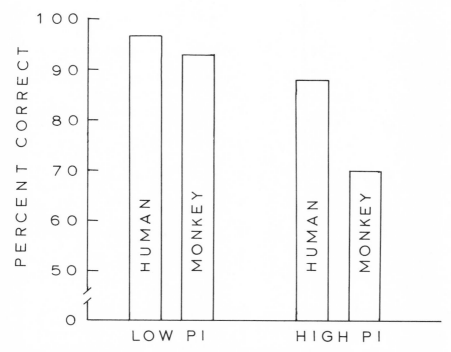

FIG. 6.3.    Percent correct recognition by a human and a rhesus monkey performing a serial-probe-recognition task under conditions of either low or high PI.

Table 6.2 shows that the relatively low level of accuracy in our high-PI condition is similar to that obtained by others studying list memory in monkeys and pigeons (Devine & Jones, 1975; Eddy, 1973; Gaffan, 1977; MacPhail, 1980; Roberts & Kraemer, 1981). The procedures used in these other experiments were comparable in many respects to our high-PI condition. Subjects were shown lists of three items to remember, and the items within individual lists were frequently repeated within a session. In fact, Gaffan and MacPhail, in their studies, used a same/different procedure very similar to the one we used in our research. Gaffan showed monkeys lists of three colors on a single panel followed by a single test color on the same panel. If the test color matched one of the list colors, the monkey obtained reward by pressing that panel; if it did not match any of the list colors, reward was obtained by pressing an alternative white panel. A total of six colors was used to generate the three-item lists. Accuracy of performance under these conditions was only 70% correct. In MacPhail's experiment, pigeons were shown lists of three stimuli on the center key of a three-key display, followed by a single test stimulus on the same key. The birds then indicated whether the test item was identical to

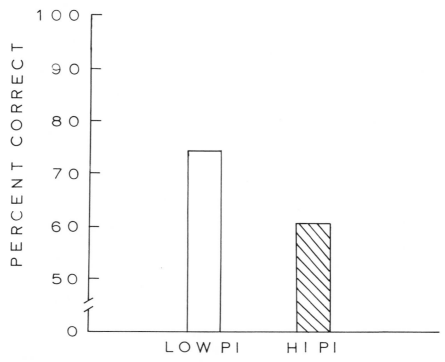

FIG. 6.4.   Mean percent correct recognition by four recently trained rhesus monkeys on a serial-probe-recognition task under conditions of either low or high PI.

one of the list items or different from all of them by pecking one of the two adjacent side keys. A total of only seven stimuli (five colors, X, and vertical lines) were used to construct the three-item lists in MacPhail's experiment. Accuracy of performance was only 54%.

The remaining three studies (Devine & Jones, 1975; Eddy, 1973; Roberts & Kraemer, 1981) differed from the others primarily by assessing list memory with either two- or three-alternative forced-choice recognition tests, much like those used in DMTS. In the Devine and Jones study, three-item lists were constructed from a pool of 10 two-dimensional stimuli. Retention tests consisted of a choice between three stimuli, one of which matched an item seen in the previous list. Under these conditions, monkeys chose the correct test item on only 65% of all trials (chance = 33%). Roberts and Kraemer (1981) also used an 10-item pool of stimuli to construct lists of one, three, or six items which were presented in random order to monkeys. The items in each list appeared on a center key of a three-key display, and the final item was followed by two test stimuli on the adjacent side keys. Accuracy of performance on the three-item lists (the condition used for comparison here) was only 62%.

The dramatic difference in performance between conditions in which list

TABLE 6.2
Three-item List Performance from Six Experiments Where
Proactive Interference Was High and from One Where It Was Low

| Proactive Interference | Experiment | Subject(s) | List Length | Percent Correct |
|---|---|---|---|---|
| High PI | Gaffan (1977) | Rhesus Monkeys | 3 (Exp 1, phase 8) | 70% |
| | MacPhail (1980) | Pigeons | 3 (Exp V) | 54% |
| | Devine & Jones (1975) | Rhesus Monkeys | 3 (3 comparisons) | 65% |
| | Eddy (1973) | Stumptail Macaques | 3 (Exp 1) | 72% |
| | Roberts & Kraemer (1981) | Squirrel Monkeys | 3 (Exp 1) | 62% |
| | Sands & Wright (1980a, b) | Rhesus Monkey | 3 (High PI) | 70% |
| Low PI | Sands & Wright (1980a, b) | Rhesus Monkey | 3 (Low PI) | 93% |

items were repeated within a session versus trial-unique is a clear example of the detrimental effect of PI in multiple-item memory. The finding that the effect appears in different species, with different retention-test procedures, and with various stimuli simply illustrates the pervasiveness of this phenomenon.

In their SPR study, Roberts and Kraemer (1981) noted that accuracy of remembering the final item in both three- and six-item lists was consistently lower than the corresponding accuracy for the individual items in single-item lists (see Figure 6.5). They interpreted this decrement as an effect of within-list (intratrial) PI: The initial items of the longer lists interfered with remembering later items. This type of analysis follows the typical PI explanation offered for many human-memory experiments (Foucault, 1928; Postman, 1969). Without disputing this interpretation, it seems somewhat curious that three of the four monkeys were *less* accurate in remembering the last item in the three-item lists than in the six-item lists. From a strictly within-list PI point of view, one would expect that this effect should be reversed.

Still, the main point is that within-list PI in the Roberts and Kraemer (1981) study was probably quite small relative to the larger background level of PI created by repeating items: repeated-item PI effect. Consider, for example, that the monkeys in the Roberts and Kraemer study averaged no better than 76% correct when only single items were presented for remembering. Corresponding accuracies for the final item in the three- and six-items lists were noticeably lower, averaging about 65% and 71% correct, respectively. (The latter figures supposedly represent the decrement due to within-list PI.) Now compare these figures with those obtained from the monkey we trained with trial-unique

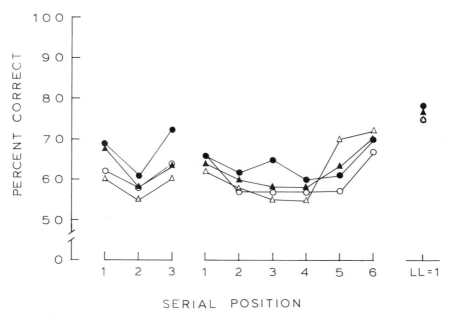

FIG. 6.5. Mean percent correct choice responses by squirrel monkeys as a function of the serial position of the tested sample item in a multiple-item delayed matching-to-sample task (The separate functions represent data from individual monkeys, from Roberts and Kraemer, 1981).

stimuli (Sands & Wright, 1980a); It correctly remembered the final item in three-item lists 96% of the time. The corresponding figure for lists of both 10- and 20-items was 88%. Although procedural differences between our experiment and Roberts and Kraemer's work might explain these performance differences, the 20% difference in accuracy between our trial-unique condition and the repeated-item condition used by Roberts and Kraemer compares very favorably with the high- to low-PI decrement of 23% we observed when training conditions were identical (cf. Table 6.2). Thus, we believe that repeated-item PI is responsible for these differences, and that it was a much larger factor in the Roberts and Kraemer experiment than the within-list PI seen directly in their data.

The apparent greater repeated-item PI effect than the within-list PI effect raises the possibility that maybe within-list PI could be better studied if repeated-item PI were eliminated. We have conducted an experiment in which it was possible to examine within-list PI when items were infrequently repeated within a session. In this experiment (Sands, 1979), list lengths of one, two, three, four, five, and six were presented in random order to the monkey during each session. The items within each list were drawn from a 211-item pool. The serial position data from this experiment are shown in Figure 6.6. It is clear

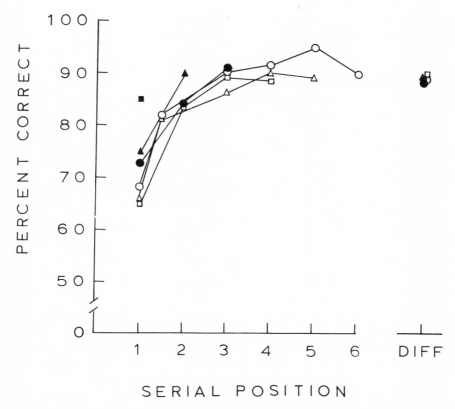

FIG. 6.6.    Percent correct recognition by a rhesus monkey as a function of serial position of the tested item in a serial-probe recognition task involving list lengths varying from one to six items within each session (Performance with each list length is plotted separately: LL1—■; LL2— ▲; LL3—●; LL4—□; LL5—△; LL6—○).

from performance on the last item of each list that there was *no* within-list PI. Accuracy of retention for the final item in lists of lengths 2–6 (LL 2–6) was no worse than for the individual item in the single-item lists (LL 1). If anything, the reverse is true.

Why was within-list PI absent in our experiment, but clearly evident in the Roberts and Kraemer (1981) experiment? One possibility might be that the different retention test procedures used in these experiments (yes/no [Sands, 1979] versus two-alternative forced-choice [Roberts and Kraemer, 1981]) were differentially sensitive to picking up a within-list PI effect. With forced-choice recognition, confusion could potentially occur on every trial if the incorrect alternative resembled one of the list items. Furthermore, since some of the stimuli used by Roberts and Kraemer were quite similar to one another (see their Figure 1), generalization between an incorrect alternative and a nontested list item was probably an important factor contributing to the frequent selection of the incorrect choice. In our yes/no procedure, interference could only be a factor on different trials (viz. subjects might respond "same" because the different probe resembled one of the list items), and a small one at that because the stimuli were multidimensional.

The analysis of within-list PI in terms of item similarity is consistent with the analysis of intratrial PI in DMTS. Recall that the intratrial PI effect was shown to be specific to presentation of the alternative (incorrect) choice stimulus as the presample event. Irrelevant stimuli, those whose characteristics supposedly do not resemble either choice alternative in DMTS, do not produce intratrial PI (Medin, 1980; Zentall & Hogan, 1974, 1977). The analysis offered here is also consistent with the finding that PI in human list memory is directly related to the similarity between the characteristics of the potentially interfering items and the items to be remembered (Underwood, 1983; Wickens et al., 1963).

## Reduction of PI and Memory for Lists of Six or More Items

Three studies are considered in this section, one using a chimpanzee (Buchanan, Gill & Braggio, 1981), another using a rhesus monkey (Sands & Wright, 1980 a,b), and a third using a dolphin (Thompson & Herman, 1977). They all share the characteristic that at least one experiment used list lengths of six or more items, and item pools were no smaller than 35 items. Under these conditions, subjects performed much more accurately than their counterparts in experiments involving smaller pool sizes and more frequent item repetition.

Buchanan, Gill and Braggio (1981) trained a chimpanzee (Lana) in a list-memory study involving 35 different Yerkish symbols (15 symbols denoted types of food or drink, 12 denoted different objects, and 8 stood for different colors). A list of items was presented to Lana by lighting symbols successively on a display board directly above a 112-pushbutton response matrix. Following presentation of the list, the chimp could obtain reward by pressing all of the buttons on the response matrix which corresponded to the items seen in the list. This test procedure was, in essence, a multiple recognition task involving a sequence of $n$ alternative forced choices. The omission of any list item or the selection of any nonlist item during the test cancelled reward. Lana performed at about 80% accuracy with four- and five-item lists and at about 70% accuracy with longer lists. Figure 6.7 shows Lana's accuracy as a function of serial position of the items in an eight-item list, the longest list used in this experiment. List-memory performance is impressive both in terms of accuracy and in terms of the serial position effects, which resemble those seen with humans. Although the 35 items of the pool were undoubtedly reused during the daily sessions, the frequency of reuse is unclear. However, Buchanan et al. were

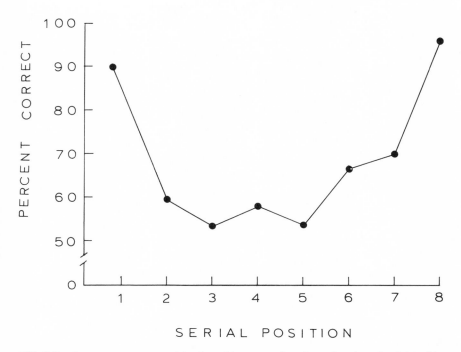

FIG. 6.7.    Percent correct recognition by a chimpanzee (Lana) as a function of serial position in a multiple-item recognition task (After Buchanan, Gill & Braggio, 1981).

aware of the problems arising from repeating items within a session and took steps to minimize the interference such repetition would cause: "During the entire series of trials, each successive word list was from a different semantic category in order to reduce proactive interference from one trial to another" (p. 652). The care they used in minimizing PI was probably instrumental in obtaining good list-memory performance with Lana.

Sands and Wright (1980a) trained a rhesus monkey on an SPR task using a pool of 211 different color slides. The monkey's performance with 10- and 20-item lists is shown in Figure 6.8. The serial-position function for the 10-item list is somewhat smoother than that for the 20-item list because more trials were conducted with the former. Nonetheless, average performance was greater than 80% correct for both list lengths: 88% with 10 items, and 83% with 20 items. Furthermore, both primacy and recency effects were evident for each list. The level and the nature of this list-memory performance is typical of four other monkeys which we have subsequently trained in this task. The monkey whose performance is shown in Figure 6.8 has since improved its accuracy to better than 90% correct with 10-item lists. We believe that the key to this highly accurate performance with long list lengths is the large item pool of distinctly different stimuli and the absence of item repetition within a session.

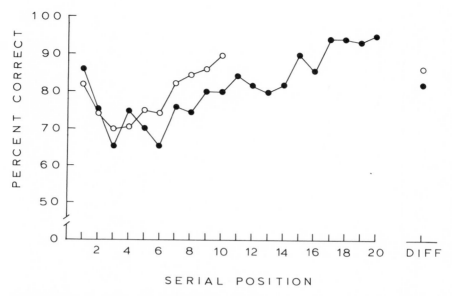

FIG. 6.8.    Percent correct recognition by a rhesus monkey as a function of serial position of the tested item in a serial-probe-recognition task with 10- and 20-item lists.

Thompson and Herman (1977) trained a dolphin in an SPR task using auditory stimuli. Lists of sounds were presented through one underwater speaker, and the probe sound through a different underwater speaker. The dolphin nosed one paddle to indicate "same" (the probe sound was contained in the just-heard list), and another paddle to indicate "different" (the probe was not contained in the list). Lists lengths of one, two, three, four, and six items were intermixed within a session. Figure 6.9 demonstrates the high level of accuracy achieved by the dolphin in this task: nearly 90% correct with three-item lists and 72% with six-item lists.

The number of different sounds used in the Thompson and Herman (1977) experiment was considerable: one hundred sounds in each of six different sound classes. These experimenters also attempted to minimize PI in the following way: "All 100 sounds in a class were used before any was repeated. Absolute frequencies of the different sound in the list were separated by at least 140 jnd's. Except in the case of six sound lists, a new probe sound [probe on Different trial] was always of a different sound class than any used in the list" (p. 502).

These experiments strongly suggest that minimizing what we have called repeated-item PI is instrumental in obtaining accurate list-memory performance in animals. Obviously, this type of PI can be minimized (or perhaps eliminated) by using as many different stimuli as possible and by organizing sessions such that, when items are repeated, they are separated by many

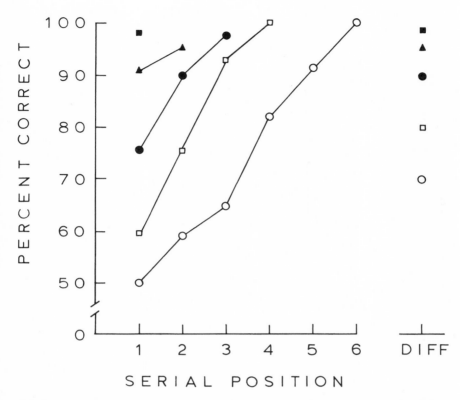

FIG. 6.9.  Percent correct recognition by a bottle-nosed dolphin as a function of serial position of the tested item in an auditory serial-probe-recognition task involving list lengths varying from one to six items within each session (Performance with each list length is plotted separately: LL1—■; LL2—▲; LL3—●; LL4—□; LL6—○; from Thompson and Herman, 1977).

intervening items. Ideally, all items would be novel but this is practically, if not theoretically, impossible. A good alternative is to use trial-unique items: items which appear only once during a daily session. Unfortunately, even this may not be possible given the various constraints on apparatus and available stimuli encountered by experimenters. The question of interest then becomes, how far does one need to separate repetitions of a stimulus in order to eliminate the PI it might cause? This question is considered in the next section.

### Extent of the Repeated-item PI Effect

Sands, (1979) investigated PI in animal list memory as a function of item separation. Interference was systematically controlled by previewing selected probe items on "different" trials, and arranging that these items were seen in a previous list. The number of intervening items separating the current probe from its prior sighting was varied from 11 to 60. The presentation rate for individual list items was also varied to examine how PI might be influenced by the amount of time separating these two events.

Forty-eight sessions, each containing 140 trials, were run with trial-unique stimuli (with the obvious exception of the explicitly manipulated interfering stimuli). Sessions were composed of 70 "same" trials and 70 "different" trials, with 10-item lists presented on each trial type. The 70 "different" trials were divided into 20 no-interference control trials (the probe had not been seen in any prior list) and 50 interference trials (the probe had been seen in a prior list). The rate of item presentation was manipulated by varying the interval between successive items within a list: interitem delays of 0.8, 2.0 and 4.0 s in that order, were run over the course of 16 sessions.

Figure 6.10 shows that performance was less accurate the fewer the number of intervening items separating the "different" probe from its sighting in a previous list. PI was still evident (10% accuracy decrement) even with 51–60 items separating these events. Slower rates of item presentation also had a debilitating effect on performance, but this effect was not specific to interference trials. Lower levels of accuracy were also produced with longer

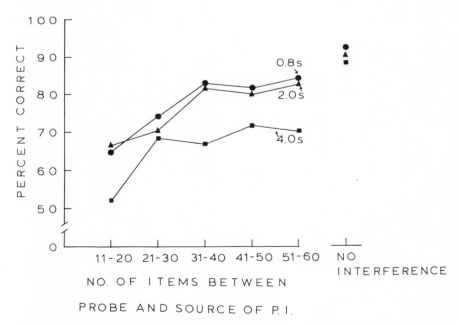

FIG. 6.10.    Percent correct recognition by a rhesus monkey on "different" trials in a serial-probe-recognition task in which the "different" probe was either seen in a prior list (with 11–60 list items intervening) or had never been seen previously (no Interference).

interitem delays on no-interference trials as well. If performance in the interference conditions are expressed as percentages of the corresponding no-interference control accuracy (not shown in the figure), presentation rate per se has little, if any, effect on PI. Thus, the number of intervening items, not the mere passage of time, seems to be the crucial factor determining the level of PI. These results are similar to those obtained by Bennett (1975) who found PI in human list memory to be a function of the number of intervening items.

In summary, this experiment demonstrated rather clearly that not only will item repetition produce PI, but that the repeated-item PI effect is felt even after 51–60 intervening items. Thus, it does not seem entirely surprising that the frequent item repetition used by many experimenters depresses assymptotic performance far below what could otherwise be obtained.

## CONCLUSION

Stimulus repetition in animal memory experiments creates much more interference than previously thought. The evidence for this statement comes primarily from studies comparing memory performances with trial-unique stimuli versus a small set of repeated stimuli; accuracy is greatly improved with trial-unique stimuli (Overman & Doty, 1980; Sands & Wright, 1980a, 1980b). Moreover, elimination of repeated-item PI may have been responsible for the animals being able to acquire SPR performance in the first place (Sands, Urcuioli, Wright, & Santiago, 1984; Wright, Sands, Santiago, & Urcuioli, 1984; Santiago & Wright, 1984; Wright, Santiago & Sands, 1984). The results from these experiments have clear and important implications for researchers studying the extent and nature of animal memory.

The mechanism(s) of the repeated-item PI effect is due, at least in part, to repetition of probe items on "different" trials. Apparently, previously seeing the different-trial probe item confuses the subject as to whether or not it was in the list being tested. Such confusion makes the subject more likely to err by responding "same." This confusion, resulting from the repeated-item PI effect, occurs even when there are as many as 60 items separating the different-trial probe item and its previous viewing (see previous section). It has yet to be resolved whether or not the repeated-item PI effect is strictly limited to different-trial probe items. Experiments need to be done where probe items are shown as previous probe items (in addition to previous list items), and where same-trial probe items are previously shown as list or probe items. Previously showing same-trial probe items is an interesting mirror-image experiment to the one on different-trial probe items (Sands, 1979). Previously showing same-trial probe items should help performance, not hinder it as with previously showing different-trial probe items. The subject may not remember that the probe was in the list (on same trials), but the "confusion" from having previously seen it should tend to increase the chances that the subject will make the correct response (but for the wrong reasons). Thus, it is proposed here that repeating same-trial probe items should increase performance, and repeating different-trial probe items should decrease performance; performances should separate in opposite directions from baseline performance.

These repeated-item PI effects are working against one another when small numbers of stimuli are repeated many times—the typical situation in learning experiments. Traditional learning or discrimination experiments generally have repeated small numbers of stimuli so that associations to individual stimuli will be "stamped in" through repetition. In recent years, conditional discrimination paradigms (e.g., MTS, SPR) have become very popular learning experiments; they are based on a relationship between pairs of stimuli (MTS) or among a set of stimuli (SPR) rather than individual stimuli. Conducting counterbalanced sessions so that every stimulus is in all the roles (e.g., MTS: sample/correct comparison/incorrect comparison) an equal number of times produces the repeated-item PI effect. However, notice that it no longer achieves the goal of strengthening a particular association (stimulus-response) in the traditional learning sense; the stimulus is the incorrect comparison as often as it is the correct one. The desire, in this situation, is to strengthen the association between a relationship (a stimulus-stimulus relationship such as identity in MTS) and the correct response. This counterbalanced procedure seldom produces evidence that pigeons learn a matching concept. Is the repeated-item PI effect somehow preventing concept learning? Perhaps interference (repeated-item PI) produces so much disruption and errors during the crucial learning stages that the more general rule is not learned. The alternative, not changing the role of the stimuli, would make the same few stimuli always the correct choices (in MTS). The pigeon might just memorize the correct choice and ignore the sample stimuli. Experiments could be conducted where stimuli are presented "trial unique" (roles unchanging), and enough stimuli used so that memorizing the correct ones would be unlikely. This situation would possibly foster development of a matching concept, properly tested as first trial performance with novel stimuli.

## REFERENCES

Bennett, R. W. Proactive interference in short-term memory: Fundamental forgetting processes. *Journal of Verbal Learning and Verbal Behavior,* 1975, *14,* 123–144.

Blough, D. S. Delayed matching in the pigeon. *Journal of the Experimental Analysis of Behavior,* 1959, *2,* 151–160.

Buchanan, J. P., Gill, T. V., & Braggio, J. T. Serial position and clustering effects in champanzee's "free recall." *Memory and Cognition,* 1981, *9,* 651–660.

Crowder, R. G. *Principles of learning and memory.* Hillsdale, NJ: Lawrence Erlbaum Associates, 1976.

D'Amato, M. R. Delayed matching and short-term memory in monkeys. In G. H. Bower (Ed.), *The psychology of learning and motivation: Advances in research and theory* (Vol. 7). NY: Academic Press, 1973.

Devine, J. V., & Jones, L. C. Matching-to-successive-samples: A multiple-unit memory task with rhesus monkeys. *Behavior Research Methods and Instrumentation*, 1975, *7*, 438–440.

Eddy, D. R. *Memory processing in Macaca Speciosa: Mental processes revealed by reaction time experiments*. Unpublished doctoral dissertation, Carnegie-Mellon University, Pittsburgh, PA, 1973.

Foucault, M. Les inhibitions internes de fixation. *Annee Psychologique*, 1928, *29*, 92–112.

Gaffan, D. Recognition memory after short retention intervals in fornix-transected monkeys. *Quarterly Journal of Experimental Psychology*, 1977, *29*, 577–588.

Gardiner, J. M., Craik, F. I. M., & Birtwistle, J. Retrieval cues and release from proactive inhibition. *Journal of Verbal Learning and Verbal Behavior*, 1972, *11*, 778–783.

Garfein, D. S. & Jacobsen, D. E. Proactive effects in short-term recognition memory. *Journal of Experimental Psychology*, 1972, *95*, 211–214.

Grant, D. S. Proactive interference in pigeon short-term memory. *Journal of Experimental Psychology: Animal Behavior Processes*, 1975, *104*, 207–220.

Grant, D. S. Intratrial proactive interference in pigeon short-term memory: Manipulation of stimulus dimension and dimensional similarity. *Learning and Motivation.* 1982, *13*, 417–433.

Grant, D. S., & Roberts, W. A. Trace interaction in pigeon short-term memory. *Journal of Experimental Psychology*, 1973, *101*, 21–29.

Herman, L. M. Interference and auditory short-term memory in the bottle-nosed dolphin. *Animal Learning and Behavior*, 1975, *3*, 43–48.

Herman, L. M. & Thompson, R. K. R. Symbolic, identity, probe delayed matching of sounds by the bottle-nosed dolphin. *Animal Learning and Behavior*, 1982, *10*, 22–34.

Hogan, D. E., Edwards, C. E., & Zentall, T. R. Delayed matching in the pigeon: Interference produced by the prior delayed matching trial. *Animal Learning and Behavior*, 1981, *9*, 395–400.

Honig, W. K., & Thompson, R. K. R. Retrospective and prospective processing in animal working memory. In G. H. Bower (Ed.). *The psychology of learning and motivation: Advances in research and theory* (Vol. 16). NY: Academic Press, 1982.

Jarrad, L. E. & Moise, S. L. Short-term memory in the monkey. In L. E. Jarrad (Ed.), *Cognitive processes of nonhuman primates*. NY: Academic Press, 1971.

Jarvik, M. E., Goldfarb, T. L. & Carley, J. L. Influence of interference on delayed matching in monkeys. *Journal of Experimental Psychology*, 1969, *81*, 1–6.

Keppel, G., & Underwood, B. J. Proactive inhibition in short-term retention of single items. *Journal of Verbal Learning and Verbal Behavior*, 1962, *1*, 153–161.

Loess, H. & Waugh, N. Short-term memory and intertrial interval. *Journal of Verbal Learning and Verbal Behavior*, 1967, *6*, 455–460.

MacPhail, E. M. Short-term visual recognition memory in pigeons. *Quarterly Journal of Experimental Psychology*, 1980, *32*, 531–538.

Maki, W. S., Moe, J. C., & Bierley, C. M. Short-term memory for stimuli, responses, and reinforcers. *Journal of Experimental Psychology: Animal Behavior Processes*, 1977, *3*, 156–177.

Mason, M., & Wilson, M. Temporal differentiation and recognition memory for visual stimuli in rhesus monkeys. *Journal of Experimental Psychology*, 1974, *103*, 383–390.

Medin, D. L. Proactive interference in monkeys: Delay and intersample interval effects are noncomparable. *Animal Learning and Behavior*, 1980, *8*, 553–560.

Mishkin, M., & Delacour, J. An analysis of short-term visual memory in the monkey. *Journal of Experimental Psychology: Animal Behavior Processes*, 1975, *1*, 326–334.

Moise, S. L. Proactive effects of stimuli, delays and response position during delayed matching to sample. *Animal Learning and Behavior,* 1976, *4,* 37–40.

Overman, W. H. & Doty, R. W. Prolonged visual memory in macaques and man. *Neuroscience,* 1980, *5,* 1825–1831.

Petrusic, W. M. & Dillon, R. F. Proactive interference in short-term recognition and recall memory. *Journal of Experimental Psychology,* 1972, *95,* 412–418.

Postman, L. Mechanisms of interference in forgetting. In G. A. Talland & N. C. Waugh (Eds.), *The pathology of memory.* NY: Academic Press, 1969.

Roberts, W. A. Distribution of trials and intertrial retention in delayed matching to sample with pigeons. *Journal of Experimental Psychology: Animal Behavior Processes,* 1980, *6,* 217–237.

Roberts, W. A. & Grant, D. S. Studies of short-term memory in the pigeon using the delayed matching-to-sample procedure. In D. L. Medin, W. A. Roberts, & R. T. Davis (Eds.), *Processes of animal memory.* Hillsdale, NJ: Lawrence Erlbaum Associates, 1976.

Roberts, W. A., & Kraemer, P. J. Recognition memory for lists of visual stimuli in monkeys and humans. *Animal Learning and Behavior,* 1981, *9,* 587–594.

Roberts, W. A. & Kraemer, P. J. Some observations of the effects of intertrial interval and delay on delayed matching-to-sample in pigeons. *Journal of Experimental Psychology: Animal Behavior Processes,* 1982, *8,* 342–353.

Roitblat, H. L. Codes and coding processes in pigeon short-term memory. *Animal Learning and Behavior,* 1980, *8,* 341–351.

Roitblat, H. L. & Scopatz, R. A. Sequential effects in pigeons delayed matching-to-sample performance. *Journal of Experimental Psychology: Animal Behavior Processes,* 1983, *9,* 202–221.

Sands, S. F. *Primate memory: Probe recognition performance by a rhesus monkey.* Unpublished doctoral dissertation, University of Texas Graduate School of Biomedical Sciences, Houston, TX, 1979.

Sands, S. F., Urcuioli, P. J., Wright, A. A., & Santiago, H. C. Serial position effects and rehearsal in primate visual memory. In H. L. Roitblat, T. G. Bever, & H. S. Terrace (Eds.), *Animal cognition.* Hillsdale, NJ: Lawrence Erlbaum Associates, 1984.

Sands, S. F. & Wright, A. A. Serial probe recognition performance by a rhesus monkey and a human with 10- and 20-item lists. *Journal of Experimental Psychology; Animal Behavior Processes,* 1980(a), *6,* 386–396.

Sands, S. F., & Wright, A. A. Primate memory: Retention of serial list items by a rhesus monkey. *Science,* 1980(b), *209,* 938–940.

Sands, S. F., & Wright, A. A. Human and monkey pictorial memory scanning. *Science,* 1982(a), *216,* 1333–1334.

Santiago, H. C., & Wright, A. A. Pigeon memory: *Same/Different* concept learning, serial probe recognition acquisition, and probe delay effects on the serial-position function. *Journal of Experimental Psychology: Animal Behavior Processes,* 1984, *10,* 498–512.

Thompson, R. K. R., & Herman, L. M. Memory for lists of sounds by the bottle-nosed dolphin: Convergence of memory processes with humans? *Science,* 1977, *195,* 501–503.

Thompson, R. K. R., & Herman, L. M. Auditory delayed discriminations by the dolphin: Nonequivalence with delayed matching performance. *Animal Learning and Behavior,* 1981, *9,* 9–15.

Underwood, B. J. "Conceptual" similarity and cumulative proactive inhibition. *Journal of Experimental Psychology: Learning, Memory, and Cognition,* 1983, *9,* 456–461.

Watkins, O. G., & Watkins, M. J. Buildup of proactive inhibition as a cue-overload effect. *Journal of Experimental Psychology: Human Learning and Memory,* 1975, *104,* 442–452.

Wickens, D. D., Born, D. G., & Allen, C. K. Proactive inhibition and item similarity in short-term memory. *Journal of Verbal Learning and Verbal Behavior,* 1963, *2,* 40–445.

Wilkie, D. M., & Spetch, M. L. Pigeons' delayed matching to sample errors are not always due to forgetting. *Behavior Analysis Letters*, 1981, *1*, 317–323.

Williams, B. A. The effects of intertrial interval on discrimination reversal in the pigeon. *Psychonomic Science*, 1971, *23*, 241–243.

Worsham, R. W. Temporal discrimination factors in the delayed matching-to-sample task in monkeys. *Animal Learning and Behavior*, 1975, *3*, 93–97.

Wright, A. A., Santiago, H. C., & Sands, S. F. Monkey memory: *Same/Different* concept learning, serial probe acquisition, and probe delay effects. *Journal of Experimental Psychology: Animal Behavior Processes*, 1984, *10*, 513–529.

Wright, A. A., Santiago, H. C., Urcuioli, P. J., & Sands, S. F. Pigeon and monkey serial probe recognition: Acquisition, strategies, and serial position effects. In H. L. Roitblat, T. S. Bever, & H. S. Terrace (Eds.), *Animal cognition*. Hillsdale, NJ: Lawrence Erlbaum Associates, 1984.

Zentall, T. R., & Hogan, D. E. Memory in the pigeon: Proactive inhibition in a delayed matching task. *Bulletin of the Psychonomic Society*, 1974, *4*, 109–112.

Zentall, T. R., & Hogan, D. E. Short-term proactive inhibition in the pigeon. *Learning and Motivation*, 1977, *8*, 367–386.

# III THEORETICAL ISSUES

# 7  AIM: A Theory of Active and Inactive Memory

Donald F. Kendrick
*Middle Tennessee State University*

Mark E. Rilling
*Michigan State University*

In this chapter, we present a theory of the processes of short-term memory as derived from animal and human data, and then consider a variety of animal short-term memory paradigms and data within the context of the proposed theory. The objective of comparing this data across species is to determine the veracity of the evolutionary continuity of the development of cognition. This continuity suggests that differences in the mental abilities of species are quantitative, rather than qualitative. Darwin (1871/1981) considered this to be his most important contribution to the understanding of the origin and evolution of life. We believe that the task of determining the differences among the memory abilities of species is best approached within a theoretical framework of animal memory. The theory that we present here is only a first approximation based on the currently available data. Change is a necessary process of theory construction. Our theoretical views are therefore a snapshot, a static representation of what we perceive at this point in time.

Before discussing the proposed theory of memory, it is necessary to describe the major paradigm used to study short-term memory in animals: delayed matching-to-sample (DMTS). The details of this paradigm are discussed in the next section. In the following sections, an overview of the theory is presented, and empirical data from animal memory studies are analyzed in terms of the theory.

## DELAYED MATCHING-TO-SAMPLE

Delayed matching procedures have been extensively employed in investigations of animal short-term memory. Though these procedures differ

129

considerably across experiments, all are composed of four phases as shown in Figure 7.1. First, an event to-be-remembered, a sample stimulus (S), is presented. Second, a delay is placed between the event to-be-remembered and the test event (or comparison stimuli). Third, the test for memory of the to-be-remembered event is given (the presentation of the comparison stimuli). Fourth, the intertrial interval occurs, after which the cycle repeats, with a different to-be-remembered event.

In the basic procedure, the set of to-be-remembered events—the sample stimuli—is restricted to two; they are randomly ordered as to which one appears after the intertrial interval. As shown in Table 7.1, there are two methods for presenting test events: a choice procedure or a go/no go procedure. In the choice procedure, shown in the top portion of Table 7.1 and Figure 7.1, two stimuli are presented after the delay-interval, and a choice response to the correct one (the stimulus that matches the most recently presented sample stimulus) is reinforced. A choice response to the incorrect stimulus (the one that does not match the sample stimulus) initiates the intertrial interval without reinforcement. In the go/no go procedure, described in the bottom portion of Table 7.1, a single test or comparison stimulus is presented. Responding is reinforced if the response matches the preceding sample stimulus, and is not reinforced if it does not match the preceding sample stimulus.

Short-term memory is considered to be involved in these procedures because correct responding depends on the memory of the sample stimulus which is not physically present at the time of the test, and because the interval between the sample and the test stimuli is often very short (1 to 15s). The sample stimulus must, in some fashion, extend its influence across time to the occurrence of the test stimuli.

## The Nature of Representations

It is commonly accepted that sample stimuli may extend their influence across the delay in three different ways (e.g., Riley, Cook & Lamb, 1981). First, pigeons may form sample-specific mediating behaviors (Zentall, Hogan, Howard & Moore, 1978) that serve to bridge the delays between the sample and test stimuli. For example, a red sample may generate a higher rate of responding than a green sample; the pigeon chooses the red test stimulus if responding at a high rate when the test stimuli are presented. Second, pigeons may code the sample as an instruction of which test stimulus to choose on the upcoming test. For example, a red sample stimulus is coded as the instruction to "choose the red test stimulus." Third, attributes of the sample may form an image representation. For example, the sample stimulus may be represented as an image of "redness" or "greenness."

There is no reason to assume that pigeons employ one strategy to the exclusion of others. All strategies may be employed by the same subject within

SAMPLE     DELAY    COMPARISON RFT/ITI
   ON     INTERVAL   STIMULI

FIG. 7.1.   The four components of a delayed matching-to-sample trial.

TABLE 7.1
Two Delayed Matching-To-Sample Procedures

| Sample Stimulus (Center Key) | Delay | Two-Choice Test | |
|---|---|---|---|
| | | Left Key | Right Key |
| Red | ——— | Red (+) | Green (−) |
| Red | ——— | Green (−) | Red (+) |
| Green | ——— | Red (−) | Green (+) |
| Green | ——— | Green (+) | Red (−) |

*The Two-Choice Procedure* is at the top.

*The Successive (Go/No Go) Procedure*

| Sample Stimulus (Center Key) | Delay | Test (Center Key) |
|---|---|---|
| Red | ——— | Red (+) |
| Red | ——— | Green (−) |
| Green | ——— | Green (+) |
| Green | ——— | Red (−) |

a session or across sessions. Furthermore, it may be that pigeons combine these encoding strategies: a red stimulus may generate an image of redness, elicit a high rate of pecking, and activate memory of the appropriate behavior for the anticipated test stimuli. In short, stimuli may be coded in a multiple-attribute format based not only on the physical dimension of the stimulus, but also on previous experiences with that stimulus and other similar stimuli. One avenue of research is to determine the encoding strategies pigeons use and the external environmental contingencies that promotes them.

## ACTIVE AND INACTIVE MEMORIES: AN OVERVIEW OF THE MEMORY SYSTEM

AIM is the acronym for active and inactive memories, a two-state model of a memory system. The major features of AIM are shown in Figure 7.2. Although active memory is shown as a box, separate from inactive memory, this is not intended to indicate "two-stores." Indeed, the explicit notion of this model is that there is a single memory store, or buffer, in which all experiences are transformed into memorial representations. These memories may then exist in one of two states. The usual state of the bulk of memories is inactive; this corresponds to the traditional concept of the long-term memory store. Memories may also be in an active state, by which it is meant that they occupy consciousness. This state corresponds to the traditional concept of short-term memory (e.g., Atkinson & Shiffren, 1968).

Figure 7.2 shows the general process of the transformation of stimulus inputs (S) into responses (R) as envisioned by AIM. Changes in stimulus input

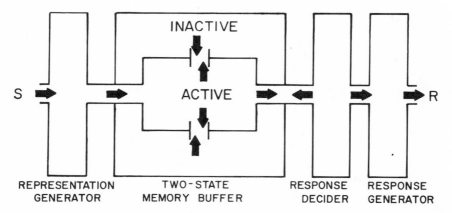

FIG. 7.2.   The animal memory system showing inactive memory and active memory as two states or a single memory buffer.

generate representations of the stimuli that are initially active, but are quickly "stored" in the inactive memory buffer. Inactive memories may also become activated, i.e. retrieved. Active memories or representations are the basis of response decisions. AIM assumes that active stimulus information generates response decisions, which in turn generates responses. Once a response decision is reached and a response generated, active memories are inhibited. That is, active representations generate one and only one response decision. Reactivation is required for each new response decision.

## Inactive Memory

A single buffer containing all memories as distinct "frames" is proposed here. Recent events generate new memories that are added to the "conveyor belt" memory buffer so that temporal information is retained by the ordinal position of the memory in the buffer. Memories are encoded one at a time and stored in temporal order. Retrieval of a memory requires a backward scanning, self-terminating, search process. Memories selected by such a search process are "projected" into an active memory state. This notion has many counterparts in the human-memory literature (e.g., Anderson, 1973; Baddeley & Hitch, 1974), but in this case resembles the "conveyor belt" model proposed by Murdock (1974), and Murdock and Anderson (1975).

The criteria for generating a memorial representation are assumed to be similar to the principles of stimulus generalization. A new memory "frame" is generated when stimulus input changes sufficiently from the immediately previous input. Small changes in stimulus input are not sufficient to generate a new representation. Memorial representations are generated automatically when a stimulus is detected. Automatic processes do not require conscious effort and control (see Hasher & Zacks, 1979 for discussions of automatic and effortful processing).

## Active Memory

Currently active stimulus representations control behavior. When environmental conditions change, new response decisions are required and the active stimulus representations reflect the changed environmental conditions. It is assumed that changes in stimulus input generate stimulus representations of the new stimuli, which are active. Newly generated representations are favored for activation. These new representations, once active, effect automatic scanning of the inactive memory buffer. This scanning process is responsible for selecting previously encoded information for activation. Thus, current stimulus representations and past representations may be simultaneously active.

It is also assumed in the AIM model that the number of memories that can be simultaneously active is limited (e.g., Atkinson & Shiffren, 1968; Wagner, 1981). The maximum number of simultaneously active representations, or chunks of information, may be one way to conceptualize species differences in memory processing. Denny (1980) has proposed such a notion in terms of task complexity: "Complexity equals the numbers of conjunctively relevant stimuli needed to elicit the target response; these stimuli can occur simultaneously or over a brief period of time and, depending on the nature of the target response, they can be held in immediate memory or be presented concurrently (p. 249)." Denny further limits the number of stimuli that can be held in immediate memory in terms of span. For example, the delayed matching task requires a span of two, and memory of a string of six digits in proper order requires a span of six. Comparatively, evidence suggests that pigeons have a span of two or three, monkeys a span of four, and human span has been estimated at five to seven (plus or minus two).

The implication of span for our discussion is that it may characterize quantitative differences in memory processing across species, thus allowing for a general theory of memory. Span also limits the number of simultaneously active representations. A limited-capacity memory system requires processes of activation and deactivation of representations.

Three processes of activating stimulus representations are recognized (see Wagner, 1981). First, it is assumed that with the presentation of an environmental event, a representation of that stimulus is activated. This is self-generated activation. If the number of active representations is at the maximum, this will result in the deactivation of one or all of the currently active representations. That is, a stimulus change large enough to generate a distinct memorial representation commands attention to the expense of the previous representations.

Second, whenever a representation is self-activated, a memory search is initiated for previous instances of the active representation and these memories are activated. This is similarity-generated activation. It is assumed that the scanner is searching for previous instances of the same event. The "rule" for identifying a representation as a past instance of a current event is a probability function based on similarity. The more similar the representation to the current stimulus, the more likely it will be identified as a past instance. Thus, past events that are similar to current events, as well as previous instances of the current event, are automatically recalled. (This assumption of similarity-generated activation also allows for the automatic retrieval of frequency information; cf. Hasher & Zacks, 1979).

Third, active representations tend to effect the activation of associated representations. Associated representations are defined by the principles of associative learning. Thus, previously unrelated events may, through learning, come to activate representations of one another upon presentation of one of them.

These three activations processes—self-generated, similarity-generated, and associatively-generated activation—are proposed to be search rules that the scanner follows in searching the memory store. Recall that memories are stored in a conveyor-belt fashion retaining temporal order. Thus, rather than assuming that memories are arranged in a network fashion, or hierarchially, the AIM model assumes that only the search processes follow a network-like pattern. Moreover, the search rules are automatic decision rules, i.e., they control the direction of the flow of spreading activation. By these three activation processes, the "stream of consciousness" is maintained.

## BASIC PHENOMENA

In this section, four basic phenomena—the memory trace, proactive and retroactive interference, directed forgetting, and primacy and recency effects—are discussed. The basic results in these areas are shown to be compatible with the predictions of the AIM model, and in some cases apparent contradictory empirical findings are resolved by viewing the findings in the light of AIM. The purpose of this section is to show: (1) that AIM integrates a wide range of short-term memory phenomena under one theoretical umbrella; and (2) that AIM generates predictions for future research.

### The Memory Trace

*Stimulus Duration.* It is well-established that the longer a stimulus is present, the better it is remembered. This has been explained in terms of rehearsal (Roberts & Grant, 1978a). Rehearsal refers to the maintained activation of a stimulus representation. Rehearsing a stimulus increases the probability that it will be stored in long-term memory and that it will be active at the time of the immediate memory test. Rehearsal has also been proposed as the process responsible for accurate delayed matching in animals. For example, the sample stimulus may be rehearsed until the time of the test. The longer the sample stimulus is present, the stronger the memory trace.

The AIM model offers a different interpretation of the stimulus-duration effect. The longer the sample stimulus is available, the greater the probability that attention will shift away from it to other environmental features or other memories. When attention is shifted back to the sample stimulus, a second representation of the same sample stimulus is generated. We have already stated that a new memory "frame" is generated whenever the current stimulus input differs significantly from the immediately preceding input. Shifts in attention alter the stimulus input and thus generate additional memory frames in the buffer. The basic idea involves the number of memory frames that contain a target memory relative to the number that do not. Thus, if one frame of four contains the target memory, activation at the time of the test is more likely than if one in five do.

Figure 7.3 shows the contents of the memory buffer for long and short stimulus durations and the probabilities of a successful activation of the target memory as a function of the increase in memory frames scanned. With more frames of the same stimulus available, the probability that the scanner will select the target memory is enhanced. Thus, longer stimulus durations enhance memory performance, because longer durations are more likely to result in more representations of the same stimulus to be encoded into the buffer.

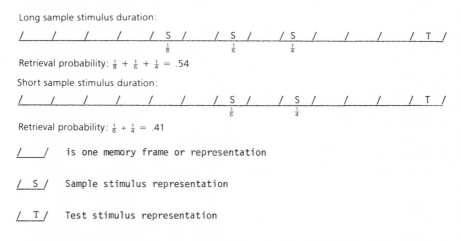

Long sample stimulus duration:

Retrieval probability: $\frac{1}{8} + \frac{1}{6} + \frac{1}{4} = .54$

Short sample stimulus duration:

Retrieval probability: $\frac{1}{6} + \frac{1}{4} = .41$

/     /     is one memory frame or representation

/  S  /    Sample stimulus representation

/  T  /    Test stimulus representation

FIG. 7.3.    The contents of the memory buffer for long and short sample stimulus durations and constant retention intervals.

NOTE: Probability of successful activation of the sample stimulus memory (S) assumes that the search set in both cases is equal. AIM assumes that probability of retrieval decreases as search set size increases. This is not considered in this figure; the decrease is assumed negligible with small search sets. T = test stimuli representation.

*Retention Interval Duration.* A second well-established finding is that long retention intervals result in poorer remembering then shorter ones. In the AIM model, this effect is due to the number of encoded representations during the retention interval between the target sample memory and the current test representation. Stimulus representations are being constantly encoded by a living attentive organism. Long retention intervals result in more additions to the memory buffer. Scanning of the memory buffer is more likely to result in the selection of the target memory the more copies there are in fewer frames.

Figure 7.4 shows the probabilities of successful activation of the target memory given long and short retention intervals. Again, the test stimuli initiate a scan of the memory buffer. With few items or memories to scan, activation of the sample memory is more probable than with many items or memories to scan.

*Repetition and Spacing.* The effects of stimulus repetition and spacing have been shown to differ between pigeons and people (Roberts, 1972). The apparent difference is, we believe, a matter of misinterpretation of the data, rather than a real difference in memory processes. Analyzing the data in terms of the AIM theory serves to demonstrate that the stimulus repetition and spacing effects are similar in pigeons and people.

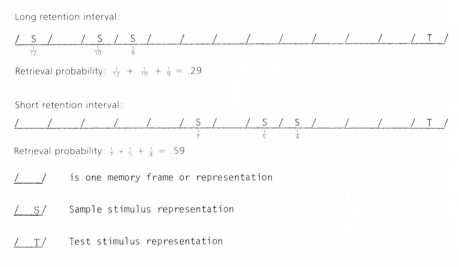

Long retention interval:

Retrieval probability: $\frac{1}{12} + \frac{1}{10} + \frac{1}{9} = .29$

Short retention interval:

Retrieval probability: $\frac{1}{7} + \frac{1}{5} + \frac{1}{4} = .59$

/___/   is one memory frame or representation

/ S/    Sample stimulus representation

/ T/    Test stimulus representation

FIG. 7.4. The contents of the memory buffer for long and short retention intervals with constant stimulus durations.

NOTE: AIM assumes that probability of retrieval decreases as search set size increases. T = test stimuli representation.

In people, spacing to-be-remembered stimuli improves memory relative to massed presentations (Bjork, 1970; Hellyer, 1962; Rundus, 1971). Roberts (1972), using the two-choice delayed matching procedure, found that pigeons performed better when multiple presentations of the sample stimulus were massed than when they were spaced. He interpreted the results as support for the trace strength and decay theory, and proposed that the memory processes in man and bird differed.

However, when Roberts' (1972) results are viewed in the context of the AIM model, they are found compatible with the findings in humans. Figure 7.5 shows the contents of the memory buffer in the two procedures of the form used by Roberts. A simplifying assumption used here is that each second of time generates a new memory frame, or representation. Thus, a 3-s delay interval is equal to three memory frames in the memory buffer. Although Roberts defined the duration of the sample stimuli in terms of the number of pecks during each, we assume each repetition generated a memorial representation.

As can be seen in Figure 7.5, more representations of the sample stimulus are generated in less time in the massed condition than in the spaced condition. This increases the probability of successful retrieval of the sample memory in the massed condition relative to the spaced condition. In contrast, the procedures involving human subjects (e.g., Bjork, 1970; Hellyer, 1962), when analyzed according to AIM, show that more representations of the target information are generated in less time in the spaced condition than in the massed condition. The critical variable for remembering, according to the

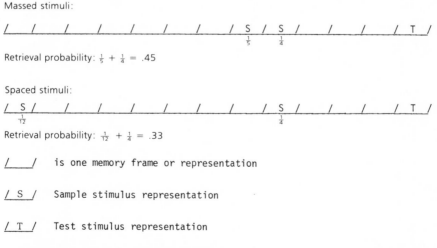

FIG. 7.5.   The contents of the memory buffer under spaced and massed conditions based on the procedures of Roberts (1972).

NOTE: AIM assumes that probability of retrieval decreases as search set size increases. T = test stimuli representations.

AIM theory, is the proportion of stimulus representations to the duration of the retention interval. The time interval between repetitions is critical. Roberts (1972) used time intervals equal to the retention interval; interstimulus intervals shorter than the retention intervals are typical in human studies and lead to the opposite prediction and result.

In the human procedures, we have the case in which massed presentations (e.g., long duration) generated few identical or similar representations of the to-be-remembered stimulus item, and spaced presentations generated many identical or similar representations of the to-be-remembered stimulus item. Thus, the probability of retrieving the target memory is better in the case where there are many representations of it, than in the case where only a few are available.

Roberts (1972) also presented stimuli in "spaced-massed" and "massed-spaced" procedures. That is, spaced presentations of the sample were followed by massed presentations, and vice versa. Humans show equal performance in both of these conditions (Bjork & Abramowitz, 1968). Pigeons are better in the spaced-massed condition than the massed-spaced condition. Again, however, these results are exactly what is predicted when the AIM model is applied to the procedures. In the human procedures, both conditions generate an equal number of representations of the target memories in an equal amount of space in the memory buffer. In the pigeon procedure, the spaced-massed condition generates more representations of the target memory in the latter portion of the memory buffer than the massed-spaced condition. As shown in Figure 7.6, the

Spaced-massed:

/ S /    /    / S /    /    /    / S / S / S /    /    / T /
$\frac{1}{12}$          $\frac{1}{9}$              $\frac{1}{5}$  $\frac{1}{4}$  $\frac{1}{3}$

Retrieval probability: $\frac{1}{12} + \frac{1}{9} + \frac{1}{5} + \frac{1}{4} + \frac{1}{3} = .97$

Mass-spaced:

/ S / S / S /    /    /    / S /    /    / S /    /    / T /
$\frac{1}{12}$ $\frac{1}{11}$ $\frac{1}{10}$              $\frac{1}{6}$              $\frac{1}{3}$

Retrieval probability: $\frac{1}{12} + \frac{1}{11} + \frac{1}{10} + \frac{1}{6} + \frac{1}{3} = .77$

/    /    is one memory frame or representation

/ S /    Sample stimulus representation

/ T /    Test stimulus representation

FIG. 7.6.  The contents of the memory buffer under spaced-massed and massed-spaced conditions based on the procedure of Roberts (1972).

NOTE: AIM assumes that the probability of retrieval decreases as search set size increases. T = test stimuli representations.

probability of a successful activation is greater in the spaced-massed condition
(.97) then in the massed-spaced condition (.77). AIM thus accounts for the
repetition and spacing data of humans and animals by showing that procedural
differences account for the differences in the data.

## Proactive and Retroactive Interference

The AIM model is based on the notion of contemporary memory theories in
which the distinction between active and inactive memory replaces the two-
stores concept of short-term and long-term memories (Lewis, 1979; Shiffrin &
Schneider, 1977).

According to AIM, proactive interference is due to a retrieval failure of the
target memory. This failure is caused by the activation of a memory other than
the target memory, which occurs when the automatic processes of similarity-
generated activation and associatively-generated activation come into play.
Moreover, the memories thus activated compete for the limited capacity
available for active memories. In brief, proactive interference is due to experimental parameters that: (1) increase the probability of activation of previous
occurrences of the test items, rather than the target memory; and (2) decrease
the probability of activation of the target memory.

A common interference paradigm is to present both sample stimuli in
succession (e.g. Nelson & Wasserman, 1978). Whether this procedure is
defined as proactive or retroactive depends on the experimenter-defined target
sample memory. If the first stimulus is the target, then the second may
retroactively inhibit or facilitate memory of the first. If the second is the target,
then the first may proactively inhibit or facilitate memory of the second. It has
been demonstrated that when the two are the same, matching accuracy is
improved, relative to single sample stimulus trials (Grant & Roberts, 1973;
Grant, 1975; Nelson & Wasserman, 1978; Roberts, 1972; Roberts & Grant,
1978b).

These effects are compatible with AIM, because the most recent stimulus is
the most likely to be active at the time of the test. The limited-activation rule
favors the most recent stimulus and, when a memory scan is initiated by the test
stimuli, the most recent relevant stimulus is more likely to be activated. Recall
that the scan self-terminates when a "satisfactory" representation—i.e., one
identified as the current sample stimulus—is activated.

Another source of proactive interference has been shown to be from the
previous trial of a delayed-matching session (Grant, 1981a; Roberts, 1980).
These studies (and others, e.g., Reynolds & Medin, 1981) have shown that
components of the test phase of the previous trial is the primary source of
proactive interference in delayed matching. AIM makes the claim that, whenever current stimulus conditions require a response decision, a memory scan
for past instances of the same stimuli is initiated. In delayed matching, this
means that the current test stimuli initiate a search for past memories involving

those stimuli; when located, they are activated and thus influence current responding. The current test stimuli also effect associatively–generated activation of the sample stimulus of the current trial. There is a competition for activation between the memory of the previous trial's test stimuli and the memory of the current sample stimulus.

Only when the association between the current test stimuli and the sample stimuli is strong enough, and conditions are arranged to increase the difficulty of the scan back to the previous trial (e.g., long intertrial interval), or to enhance the probability of a successful selection of the sample memory (e.g., long duration sample stimuli), will the sample memory control test responding.

Retroactive interference of short-term memory has been studied in the delayed matching paradigm by interpolating stimuli into the delay interval. A variety of different stimuli and events have been used as interfering stimuli including localized colored lights (Jarvik, Goldfarb, & Carley, 1969; Grant and Roberts, 1976; Kendrick & Rilling, 1984), ambient illumination (Cook, 1980; D'Amato, 1973; Kendrick, Tranberg & Rilling, 1981; Roberts & Grant, 1978a; Tranberg & Rilling, 1980), activity (Kendrick & Rilling, 1984; Moise, 1970), and tones (Herman & Gordon, 1974; Wilkie, Summers & Spetch, 1981). Of all of these stimuli, only three appear to interfere with delayed matching accuracy: activity, stimuli physically similar to the sample and test stimuli, and changes in ambient illumination. Tones appear to interfere with short-term memory only in dolphins.

Within the AIM framework, the effect of interpolated activity is explained by the increase in alternate stimulus representations that result from the change in behavioral orientation which produces a large change in stimulus input. Thus, the sample representation is deactivated by the activation of the new stimulus representations, which also increase the number of representations in the memory buffer and decrease the probability of successful retrieval at the time of the test.

The interfering effect of stimulus similarity is explained by the addition to the memory buffer of a representation that is "confused" with the target representation. The scanner searches the buffer for the sample representation and activates the first representation that satisfies the search criteria. The search criteria are established by similarity-generated activation and by associatively-generated activation. Empirical evidence suggests confusion on four dimensions: (1) physical similarity, e.g., color, size; (2) location, i.e., same position; (3) time of remembered occurrence; and (4) associations between stimulus sequences (Grant & Roberts, 1976; Kendrick, 1982; Medin, Reynolds & Parkinson, 1980; Wright, Urcuioli, Sands & Santiago, 1981; Zentall, 1973).

The interfering effects of ambient illumination are explained within the AIM framework in terms of the extent of the change in stimulus input. It has been shown that the degree of interference is proportional to the degree of illumination change; the greater the change the greater the interference (Roberts & Grant, 1978a). Increasing the amount of illumination increases the number of stimulus elements in the environment that are changed. The greater the stimulus change, the greater the probability that a target representation is rendered inactive by the newly generated representations, and the greater the probability of a retrieval failure due to the increased number of representations that must be scanned in search of the target representation.

The three sources of retroactive interference—activity, stimulus similarity, and ambient illumination changes—and the two sources of proactive interference are all explained by similar processes in the AIM model. AIM thus offers a view of the memory system of animals that unifies the diverse empirical findings in these areas under one theoretical umbrella.

## Directed Forgetting

Directed forgetting in animals is demonstrated when memory of stimuli that the animals are cued to forget is poor compared to memory of stimuli that they are not cued to forget. Directed forgetting has been demonstrated in several labs using a modified delayed matching-to-sample procedure (Grant, 1981b; Kendrick, Rilling & Stonebraker, 1981; Maki & Hegvik, 1980).

Figure 7.7 shows a typical directed forgetting procedure with pigeons. Sample stimuli are followed by a cue to forget the sample stimulus (F-cue), or by a cue to remember it (R-cue). The trial is terminated after the F-cue, i.e., no test stimuli are presented; thus the pigeon need not remember the sample stimulus. R-cued trials are the same as the typical delayed-matching trial described earlier, except the R-cue is presented after the sample, and prior to the test stimuli. These cues are typically 1s or less in duration.

To determine whether the pigeon is forgetting the sample stimulus as instructed by the F-cue, the test stimuli are infrequently presented following the F-cue, in violation of extensive baseline training. When presented with the test stimuli in this situation, matching accuracy is near chance. Trivial possibilities for this phenomenon have been eliminated (latency differences between R-cue trials and F-cue trials, Maki, Olsen & Rego, 1981; novelty of the test stimulus after F-cues, Stonebraker & Rilling, 1981).

It has been proposed that the R-cue maintains rehearsal of the sample representations, and that the F-cue terminates rehearsal (Grant, 1981b; Kendrick, Rilling & Stonebraker, 1981; Maki, Olsen & Rego, 1981; Stonebraker, Rilling & Kendrick, 1981). Retrieval of the sample information effected by test stimuli has also been proposed (D'Amato & Worsham, 1974; Kendrick, Rilling & Stonebraker, 1981).

AIM explains the directed forgetting phenomenon as follows. When the sample stimulus is presented, a representation is activated. The pigeon may then attend to other stimuli, in which case the sample representation is

R-cue

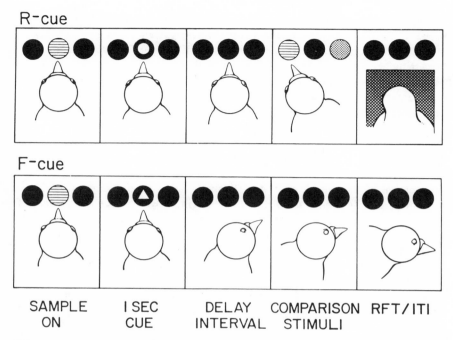

F-cue

|SAMPLE|I SEC|DELAY|COMPARISON|RFT/ITI|
|ON|CUE|INTERVAL|STIMULI||

FIG. 7.7.    The two basic trial types of directed forgetting. The R-cue trial is a typical delayed matching trial, the F-cue trial ends without test stimuli after the F-cue.

rendered inactive; then it may return to the sample stimulus, generating another active representation, and so on. Eventually, the sample stimulus is removed and the representation may remain active or become inactive, depending on whether the pigeon remains attentive or is diverted by other stimuli. The R-cue, by virtue of the anticipated test-to-come, increases the probability that the representation of the sample will remain active. The F-cue, by virtue of the anticipated omission of testing, decreases the probability that the representation will remain active.

But why do the test stimuli after F-cues fail to effect reactivation (i.e., retrieval) of the sample stimulus? Recall that stimulus changes caused by environmental changes or behavioral orientation generate new representations that deactivate old ones. In the present case, the F-cue, by virtue of signalling the end of the trial, "releases" attention to the sample representation so that attention is turned to other stimuli and other representations. The R-cue "focuses" attention on the sample stimulus representation. Thus at the time of the test, reactivation is trivial in the case of an R-cue trial, and is difficult on F-cue test trials due to the addition of representations to the memory buffer.

One variation of the basic procedure (see Figure 7.1) has been to present stimuli unrelated to the matching task after the F-cue (Grant, 1981b; Kendrick,

Rilling & Stonebraker, 1981; Maki & Hegvik, 1980). The rationale is that if the F-cue terminates rehearsal of the sample memory, then presenting a sample-independent discrimination should not alter the F-cue's function; directed forgetting will be obtained. AIM predicts that this procedure will serve to "focus" the pigeon on the task at hand, thereby reducing the number of new representations added to the buffer relative to the F-cue omission procedure. This increases the probability that the test stimuli will successfully reactivate memory of the sample; directed forgetting will not be obtained.

Kendrick et al. (1981) followed this procedure. After the retention interval of F-cued trials, a sample independent discrimination was presented. A response to the S + resulted in reinforcement, and a response to the S − did not, irrespective of the sample stimulus. R-cued trials proceeded as in the basic procedure. On the F-cued test trials, the test stimuli were presented in place of the sample-independent discrimination stimuli. Matching accuracy on these test trials did not differ from matching accuracy on the R-cued comparison trials; the pigeons did not forget the sample stimuli as shown in the basic procedure. This result fails to support the rehearsal account of directed forgetting, but does support the AIM prediction.

Maki and Hegvik (1980) demonstrated the same effect in a similar experiment. However, Grant (1981b) in a somewhat different procedure obtained exactly the opposite effect. Grant used a directed forgetting variation of the go/ no go procedure described in the introduction. On R-cue trials, a single test stimulus is presented after the retention interval, and conditional upon the sample stimulus, responding is either reinforced or not. Grant obtained the basic directed forgetting effect when the F-cue signalled the end of the trial. However, in contrast to the findings of Kendrick et al. (1981) and of Maki and Hegvik (1980), Grant also obtained forgetting when the F-cue was followed by a sample-independent "dot" stimulus. Responding during this dot stimulus was reinforced on a random one-half of the trials.

These data have implications beyond the basic directed forgetting phenomenon; a satisfactory resolution of the discrepant findings may tell us a great deal about the processes of short-term memory in pigeons. Rehearsal alone can not account for the data. Why should the F-cue terminate rehearsal of the sample representation when followed by sample-independent discriminative stimuli, but not when it is followed by a dot stimulus associated with 50% reinforced responding? Differences in reinforcement probability in the two procedures have been eliminated as a causative factor in unpublished work by Kendrick (1981). He replicated Grant's procedure and replicated the effect, but then in a second phase increased the reinforcement associated with the dot stimulus to 100%. Directed forgetting was still obtained.

There is another major difference between the two procedures; in the two-choice procedure of Kendrick et al. (1981), the F-cue was followed by a simultaneous discrimination. In Grant's (1981b) go/no go procedure, and in

Kendrick's (1981) replication, the F-cue was always followed by the same stimulus; a discrimination was not required.

A hypothesis is advanced that may account for the discrepant findings. This hypothesis is based on the AIM model as presented thus far, but is not derived from the memory processes per se. However, it has been incorporated into the general AIM framework as an integral part of the response-decision process. Before discussing the hypothesis some background information is required. Consider the scanner, or memory-search process. Memory is searched in order to make a response decision that increases the probability of a successful outcome relative to that possible without a memory search. Thus barring dramatic changes in the environment (in stimulus input), it follows that once a response decision is reached and the response initiated, a memory search should be inhibited.

How does this explain the discrepant findings in directed forgetting? In Grant's procedure, with a dot stimulus following the F-cue, a response decision can be made at the time of the F-cue. The dot stimulus provides no additional information regarding the probability of reinforcement. Once this decision is made, the memory scanner is inhibited, and when the test stimulus is presented instead of the dot stimulus, a memory search is prevented and the response decided upon at the time of the F-cue is carried out. Thus, matching accuracy is near chance. On R-cue trials, a response decision is not made until the test stimulus is presented, at which time a memory search is initiated and the sample memory activated, if it is not active when the test stimulus is presented.

In Kendrick et al's procedure, with the sample-independent simultaneous discrimination following the F-cue, a response decision can not be made until the discriminative stimuli are presented. The memory scanner is not inhibited, and when the test stimuli are presented in place of the sample-independent discrimination, a memory search is initiated. Thus, the sample memory is activated (or retrieved) and matching accuracy is good.

The obvious test of this hypothetical explanation is to replicate Grant's (1981b) procedure with one modification. Present a sample-independent successive discrimination in place of the dot stimulus. This manipulation would postpone the response decision until the time of the test, thus preventing the inhibition of the memory scanner so that when the test stimulus is presented, instead of the sample-independent stimulus, the memory buffer could be searched for the sample memory. Matching accuracy is predicted to be good, in support of our hypothesis.

Kendrick (1984) conducted this experiment based on the predictions of AIM. Three pigeons were trained on the directed forgetting procedure following Grant's (1981b) procedure; R-cued trials ended with the presentation of a single test stimulus that either matched or did not match the sample stimulus. In the first test condition (Test A), baseline F-cued trials ended with a triangle projected on the response key. Reinforcement was response-independent and occurred with a 50% probability after the 6-s triangle stimulus. In the second test condition (Test B), baseline F-cued trials ended with one of two stimuli: the triangle of Test A or a square. The triangle (S+) was associated with 100% response-independent reinforcement; the square (S−) was associated with nonreinforcement. Thus in Test B, pigeons were trained on a sample-independent successive discrimination. AIM predicts that F-cued test trial performance (red and green test stimuli after F-cues in place of the sample-independent stimuli) will be poor in Test A and that accurate matching will be obtained in Test B. Figure 7.8 shows the results of these manipulations; matching accuracy in Test B is much improved relative to Test A, and is similar to control R-cued matching accuracies. Thus, the critical variable in directed forgetting is stimulus control of response decisions. The pigeon memory system conceptualized by AIM is supported; both rehearsal and retrieval are important processes in the pigeon memory system. External stimuli may control those processes, promoting the maintenance of information (rehearsal), terminating maintenance (F-cues), and enhancing or inhibiting the activation of inactive information (retrieval).

One question remains. What constitutes a dramatic change that is great enough to disinhibit the memory scanner after a response decision? Based on Grant's (1981b) results and the current interpretations, the test stimulus in place of the dot stimulus must not be a dramatic enough change. Or is it? It may be that once a response decision is reached, attention to the current stimulus situation is reduced sufficiently to postpone stimulus input of the new (unexpected?) situation. Once it is recognized that this is a test stimulus and not the dot stimulus, the current responding must be inhibited, and the memory

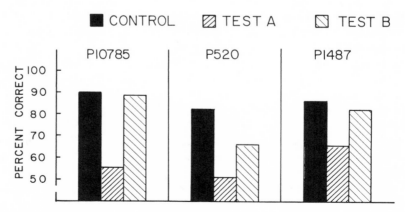

FIG. 7.8.   Matching accuracy on R-cue trials (Control) and two F-cue test conditions (Test A and Test B). Test A provides response independent reinforcement on F-cue trials, and test B requires a sample-independent discrimination on F-cue trials.

scanner disinhibited, a memory search initiated, the sample memory activated, and a new response decision generated. All of this takes time, and the stimulus is of limited duration. Assuming that the current responding continues until a new response decision is reached, the rate of responding in Grant's procedure on the test trials would indicate forgetting of the sample stimulus. Rather than attempt to define dramatic changes in stimulus input, the current data may be explained by assuming that the changes involved are dramatic enough, but require time and attention to affect response changes; time that is not available to alter the outcome of the experiment.

## Primacy and Recency Effects

Recently primacy and recency effects have been demonstrated in monkeys and pigeons (Roberts & Kraemer, 1981; Wright, Santiago, Sands & Urcuioli, 1982). Primacy refers to better memory for items at the beginning of the list than items in the middle. Recency refers to better memory for items at the end of the list than items in the middle. In humans, the serial position of an item has long been known to be an important variable in memory (see Murdock, 1960, for a review), and is evidence of a dual-process memory system. The logic is that primacy is due to retrieval from long-term memory of a representation of the beginning items, and recency is due to rehearsal in short-term memory of the last few items. The middle items suffer proactive interference from the beginning items and retroactive interference from the end items.

Serial position curves are obtained from animals by the serial probe recognition procedure; this procedure is similar to delayed matching except that a list of sample stimuli are presented, rather than just one. For example, Wright et al. (1982) constructed a slide set of lists of four color slides which were shown to pigeons and monkeys. Following presentation of the list items, and after a retention interval, a single test item was presented. If the test item matched any one of the list items, a "same" response was rewarded; if it did not match a list item, a "different" response was rewarded. Pigeons pecked response keys to the left and right of the slide screen and monkeys moved a T-lever left and right.

Wright et al. (1982) have shown that the serial position curve varies as a function of the retention interval. They tested their animals with retention intervals (which they termed probe delays) of 0, .5, 1, 2, 10, 20, and 30 s. At the shorter intervals, there was no primacy effect, but a strong recency effect. At intermediate intervals, the typical U-shaped serial position curve was obtained, demonstrating primacy and recency. At the longer intervals, there was no recency effect, but a strong primacy effect. Since a 13-year-old boy was tested and showed similar results, Wright et al. (1982) concluded that the memory system of the three species operates in similar fashion on this serial probe recognition task.

The predictions generated by AIM are consistent with the serial position and probe delay effects reviewed here. Again, the basic assumption is that presentation of each item generates a representation and that other representations are created between items. When a test item is presented the buffer is scanned and those items represented last are more easily reactivated than those items represented at the beginning of the list. This is exactly the logic used in explaining the delayed matching task.

At short-retention intervals, the last presented item has the highest probability of retrieval and the probability of retrieval of each item nearer to the beginning decreases. The problem for AIM is that this function should remain constant across all retention intervals. Indeed, there is only one way AIM can account for the serial position data of Wright et al. (1982), and that is by assuming that more copies of the first item are generated than of the other items.

Intuitively, it may be plausible that the first item generates more representations than the other items. Consider an animal during the intertrial interval. Attention is focused on something, probably something in the apparatus, the recently received reward, an itch, or some other factor. When the first item is presented, it competes for attention. The animal vacillates between the first list item and the focus of its attention prior to the presentation of the item. This vacillation of attention results in multiple representations of the first item in the memory buffer. Later in the list, the animal has settled into the task and vacillates little if any; thus constant attention to stimuli generates fewer representations than vacillating attention.

Figure 7.9 shows the probability that a representation from each serial positon is reactivated as a function of retention interval. As can be seen from this figure, the results of Wright et al. (1982) are predicted by AIM. At short delays, the last item has the highest probability of retrieval, a recency effect. At intermediate retention intervals, the U-shaped curve is obtained, and at long retention intervals, there is a primacy effect. AIM thus accurately predicts the serial-position curves across retention intervals, but only with the curious assumption about the vacillation of attention during the presentation of the first item. One way to test this assumption would be to collect eye movement data. According to AIM, eye movements should vacillate on and off the first item more than during any other items.

## CONCLUSION

In conclusion, AIM accounts for the bulk of the animal short-term memory data and provides testable predictions for future research. Moreover, AIM specifies the attentional processes that must be operative in animal short-term memory. These attentional processes are subject to empirical test.

AIM is not presented in a final or polished version, but will quite expectedly

Short retention intervals:

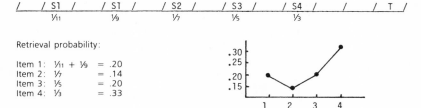

Retrieval probability:

Item 1: $1/11 + 1/9$ = .20
Item 2: $1/7$ = .14
Item 3: $1/5$ = .20
Item 4: $1/3$ = .33

Intermediate retention intervals:

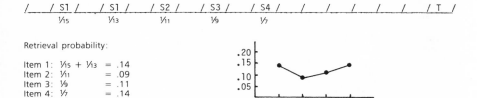

Retrieval probability:

Item 1: $1/15 + 1/13$ = .14
Item 2: $1/11$ = .09
Item 3: $1/9$ = .11
Item 4: $1/7$ = .14

Long retention intervals:

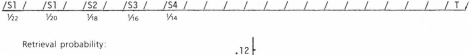

Retrieval probability:

Item 1: $1/22 + 1/20$ = .10
Item 2: $1/18$ = .06
Item 3: $1/16$ = .06
Item 4: $1/14$ = .07

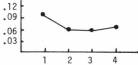

**FIG. 7.9.** The contents of the memory buffer under three retention interval durations based on the procedure of Wright et al. (1982).

NOTE: AIM assumes that the first item generates more representations than the other items. T = test item representations.

undergo transformation as more information becomes available. The data base for AIM has been the animal short-term memory literature. We hope that AIM will be extended to the long-term memory data and to the human memory literature. Such extension will necessitate changes in AIM, and in our view of memory processing in general.

## REFERENCES

Anderson, J. A. A theory for the recognition of items from short memorized lists. *Psychology Review*, 1973, *80*, 417–438.

Atkinson, R. C., & Shiffren, R. M. Human memory: A proposed system and its control processes. In K. W. Spence & J. T. Spence (Eds.), *The psychology of learning and motivation: Advances in research and theory* (Vol. 2). New York: Academic Press, 1968.

Baddeley, A. D., & Hitch, G. Working memory. In G. A. Bower (Ed.), *The psychology of learning and motivation* (Vol. 8). New York: Academic Press, 1974.

Bjork, R. A. Repetition and rehearsal mechanisms in models of short-term memory. In D. A. Norman (Ed.), *Models of human memory.* New York: Academic Press, 1970.

Bjork, R. A. Theoretical implications of directed forgetting. In A. W. Melton & E. Martin (Eds.), *Coding processes in human memory.* New York: Wiley, 1972.

Bjork, R. A., & Abramowitz, R. L. The optimality and commutativity of successive interpresentation intervals in short-term memory. Paper presented at the meeting of the Midwestern Psychological Association, Chicago, IL, May 1968.

Cook, R. G. Retroactive interference in pigeon short-term memory by a reduction in ambient illumination. *Journal of Experimental Psychology: Animal Behavior Processes,* 1980, *6,* 326–338.

D'Amato, M. R. Delayed matching and short-term memory in monkeys. In G. H. Bower (Ed.), *The psychology of learning and motivation: Advances in research and theory* (Vol. 7) New York: Academic Press, 1973.

D'Amato, M. R., & Worsham, R. W. Retrieval cues and short-term memory in capuchin monkeys. *Journal of Comparative and Physiological Psychology,* 1974, *86,* 274–282.

Darwin, C. R. *The Descent of Man, and Selection in Relation to Sex.* Princeton, NJ: Princeton University Press, 1981. (Originally published, 1871).

Denny, M. R. Complex behavior: Traditional comparative psychology. In M. R. Denny (Ed.), *Comparative psychology: An evolutionary analysis of animal behavior.* New York: Wiley, 1980.

Grant, D. S. Proactive interference in pigeon short-term memory. *Journal of Experimental Psychology: Animal Behavior Processes,* 1975, *1,* 189–206.

Grant, D. S. Intertrial interference in rat short-term memory. *Journal of Experimental Psychology: Animal Behavior Processes,* 7, 1981, 217–227. (a)

Grant, D. S. Stimulus control of information processing in pigeon short-term memory. *Learning and Motivation,* 1981, *12,* 19–39. (b)

Grant, D. S., & Roberts, W. A. Trace interaction in pigeon short-term memory. *Journal of Experimental Psychology,* 1973, *101,* 21–29.

Grant, D. S., & Roberts, W. A. Sources of retroactive inhibition in pigeon short-term memory. *Journal of Experimental Psychology: Animal Behavior Processes,* 1976, 2, 1–16.

Hasher, L., & Zacks, R. T. Automatic and effortful processes in memory. *Journal of Experimental Psychology: General,* 1979, *108,* 356–388.

Hellyer, S. Frequency of stimulus presentation and short-term retention decrement in recall. *Journal of Experimental Psychology,* 1962, *64,* 650.

Herman, L. M., & Gordon, J. A. Auditory delayed matching in the bottlenose dolphin. *Journal of the Experimental Analysis of Behavior,* 1974, *21,* 19–26.

Jarvik, M. E., Goldfarb, T. L., & Carley, J. L. Influence of interference on delayed matching in monkeys. *Journal of Experimental Psychology,* 1969, *81,* 1–6.

Kendrick, D. F. Probability of reinforcement is not a factor in directed forgetting with pigeons. Unpublished data, Middle Tennessee State University, Department of Psychology, 1981.

Kendrick, D. F. The effects of associatively-generated retrieval of stimulus information on the acquisition of a short-term memory task. PhD Dissertation, Michigan State University, 1982.

Kendrick, D. F. Procedural factors influencing directed forgetting in pigeons. Manuscript in preparation, 1984.

Kendrick, D. F., & Rilling, M. The role of interpolated stimuli in retroactive interference in pigeon short-term memory. *Animal Learning and Behavior,* 1984.

Kendrick, D. F., Rilling, M., & Stonebraker, T. B. Stimulus control of delayed matching in pigeons: Directed forgetting. *Journal of the Experimental Analysis of Behavior,* 1981, *36,* 241–251.

Kendrick, D. F., Tranberg, D. K., & Rilling, M. The effects of illumination on the acquisition of delayed matching-to-sample. *Animal Learning and Behavior,* 1981, *9,* 202–208.

Lewis, D. J. Psychobiology of active and inactive memory. *Psychological Bulletin,* 1979, *86,* 1054–1083.

Maki, W. S., & Hegvik, D. K. Directed forgetting in pigeons. *Animal Learning and Behavior,* 1980, *8,* 567–574.

Maki, W. S., Olson, D., & Rego, S. Directed forgetting in pigeons: Analysis of cue functions. *Animal Learning and Behavior,* 1981, *9,* 189–195.

Medin, D. L., Reynolds, T. J., & Parkinson, J. K. Stimulus similarity and retroactive interference and facilitation in monkey short-term memory. *Journal of Experimental Psychology: Animal Behavior Processes,* 1980, *6,* 112–125.

Moise, S. L. Short-term retention in *Macaca Speciosa* following interpolated activity during delayed matching from sample. *Journal of Comparative and Physiological Psychology,* 1970, *73,* 506–514.

Murdock, B. B., Jr. The distinctiveness of stimuli. *Psychological Review,* 1960, *67,* 61–81.

Murdock, B. B., Jr. *Human memory: Theory and data.* Potomac, MD: Lawrence Erlbaum Associates, 1974.

Murdock, B. B., & Anderson, R. E. Encoding, storage, and retrieval of item information. In R. L. Solso (Ed.), *Information processing and cognition: The Loyola symposium.* Hillsdale, NJ: Lawrence Erlbaum Associates, 1975.

Nelson, K. R., & Wasserman, E. A. Temporal factors influencing the pigeon's successive matching-to-sample performance: Sample duration, intertrial interval, and retention interval. *Journal of the Experimental Analysis of Behavior,* 1978, *30,* 153–162.

Reynolds, T. J., & Medin, P. L. Stimulus interaction and between-trials proactive interference in monkeys. *Journal of Experimental Psychology: Animal Behavior Processes,* 1981, *7,* 334–347.

Riley, D. A., Cook, R. G., & Lamb, M. R. A classification and analysis of short-term retention codes in pigeons. In G. H. Bower (Ed.), *The psychology of learning and motivation.* New York: Academic Press, 1981.

Roberts, W. A. Short-term memory in the pigeon: Effects of repetition and spacing. *Journal of Experimental Psychology.* 1972, *94,* 74–83.

Roberts, W. A. Distribution of trials and intertrial retention in delayed matching-to-sample with pigeons. *Journal of Experimental Psychology: Animal Behavior Processes,* 1980, *6,* 217–237.

Roberts, W. A., & Grant, D. S. An analysis of light induced retroactive inhibition in pigeon short-term memory. *Journal of Experimental Psychology: Animal Behavior Processes,* 1978, *4,* 219–236. (a)

Roberts, W. A., & Grant, D. S. Interaction of sample and comparison stimuli in delayed matching to sample with the pigeon. *Journal of Experimental Psychology: Animal Behavior Processes,* 1978, *4,* 68–82. (b)

Roberts, W. A., & Kramer, P. J. Recognition memory for lists of visual stimuli in monkeys and humans. *Animal Learning and Behavior,* 1981, *9,* 587–594.

Rundus, D. Analysis of rehearsal processes in free recall. *Journal of Experimental Psychology,* 1971, *89,* 63–77.

Shiffren, R. M., & Schneider, W. Controlled and automatic human information processing: II. Perceptual learning, automatic attending, and a general theory. *Psychological Review,* 1977, *84,* 127–190.

Stonebraker, T. B., & Rilling, M. Control of delayed matching-to-sample performance using directed forgetting techniques. *Animal Learning and Behavior.* 1981, *9,* 196–201.

Stonebraker, T. B., Rilling, M., & Kendrick, D. F. Time dependent effects of double cueing in directed forgetting. *Animal Learning and Behavior,* 1981, *9,* 305–394.

Tranberg, D. K., & Rilling, M. Delay interval illumination changes interfere with pigeon short-term memory. *Journal of the Experimental Analysis of Behavior,* 1980, *33,* 39–49.

Wagner, A. R. SOP: A model of automatic memory processing in animal behavior. In R. R. Miller & N. E. Spears (Eds.), *Information processing in animal behavior: memory mechanisms.* Hillsdale, NJ: Lawrence Erlbaum Associates, 1981.

Wilkie, D. M., Summers, R. J., & Spetch, M. L. Effect of delay-interval stimuli on delayed symbolic matching-to-sample in the pigeon. *Journal of the Experimental Analysis of Behavior.* 1981, *35,* 153–160.

Wright, A. A., Santiago, H. C., Sands, S. F., & Urcuioli, P. J. Pigeon and monkey serial probe recognition: Acquisition, strategies, and serial position effects. In H. L. Roitblat, T. Bever, and H. S. Terrace (Eds.), *Animal Cognition.* Hillsdale, NJ: Lawrence Erlbaum Associates, 1982.

Wright, A. A., Urcuioli, P. J., Sands, S. F., & Santiago, H. C. Interference of delayed matching-to-sample in pigeons: Effects of interpolation at different periods within a trial and stimulus similarity. *Animal Learning and Behavior,* 1981, *9,* 595–603.

Zentall, T. R. Memory in the pigeon: Retroactive inhibition in a delayed matching task. *Bulletin of the Psychonomic Society,* 1973, *1,* 126–128.

Zentall, T. R., Hogan, D. E., Howard, M. M., & Moore, B. S. Delayed matching in the pigeon: Effect on performance of sample-specific observing responses and differential delay behavior. *Learning and Motivation,* 1978, *9,* 202–218.

# 8 Delayed Alternation and Short-Term Memory in the Rat

Douglas S. Grant
*University of Alberta*

## INTRODUCTION

A "memory" is an internal representation of an event or episode encountered previously. The study of memory involves identifying both the nature of the internal representation and the processes which operate on the representation. Our understanding of memory is based on inferences drawn from overt behavior in experimental situations in which treatments either enhance retention or promote forgetting.

The data and theory discussed in this chapter concern short-term memory (STM), as distinguished from long-term memory (LTM). Studies of STM assess retention over intervals of seconds or minutes, whereas studies of LTM typically involve retention intervals of hours, days or even years. A preparation is thus required in which retention test performance is a function of memory for a previous episode, and, because our interest is STM, one in which substantial forgetting occurs over short intervals. The delayed alternation preparation meets these requirements.

A typical delayed alternation trial consists of three components; a forced turn, a retention or memory interval, and a free choice run. At the onset of a trial, the rat is forced to turn either right or left at the choice point of a T-maze and receives a reinforcer in the goal box at the end of that arm. The direction of the forced turn is determined randomly on each trial, with the restriction that over the course of a session (typically between 8 and 12 trials) half of the trials are initiated by a forced turn to the right and half by a forced turn to the left. The retention interval follows the forced turn and is typically spent in the start box. The retention test which follows this interval consists of a free choice run on which a turn in either direction is permitted. The rat is reinforced on the test run only if it turns in the direction opposite that of the forced run. To the extent that performance on the free-choice test reflects memory for information derived from the earlier forced turn, a ready technique for analyzing STM in rats is available. But in order to assert that free-choice performance in delayed alternation reflects memory, nonmemorial sources of control must be eliminated.

The issue of nonmemorial sources of control can be addressed at two levels. At one level, the issue is whether delayed alternation involves processes of representation. Roitblat (1982) viewed a representation as a code or transformation that preserves at least some information about an event or episode. The present concern, then, is whether performance on the free-choice test is controlled by a representation established or activated by a forced turn and maintained during the retention interval. To put the question conversely, is it possible that the rat might alternate accurately at the time of testing in the absence of a representation of information derived from the forced turn?

Investigators employing the delayed alternation preparation have viewed the use of odor trails as the only potentially viable source of nonrepresentational control over test responding. Specifically, it is conceivable that free-choice behavior might be controlled by an odor trail established on the initial, forced run. If this odor trail lingered, the rat might learn to avoid the arm associated with the odor on the test run. If so, delayed alternation would not involve the representation of information and would be inappropriate for the analysis of STM. Early investigators tested for use of olfactory cues by conducting initial and test runs in different mazes (Hunter, 1941) and by interchanging the maze arms between the forcing and free choice (Estes & Schoeffler, 1955). Because neither manipulation markedly reduced accuracy, alternation can hardly be based upon avoidance of odor trails.

These findings strongly suggest that delayed alternation does involve the representation of information derived from the forced run. However, to conclude that delayed alternation may be used to assess short-term retention is still premature. A second level of the issue of nonmemorial control must first be considered: is the representation in fact central? It could be argued that information from the forced run is represented in the form of a particular pattern of muscular tension (kinesthetic trace) or in the form of a maintained bodily orientation (mediating behavior). Although in both cases test performance would be controlled by information derived from the forcing, the failure of this information to be represented centrally would render the preparation useless in the analysis of STM (see Roitblat, 1982 for further discussion).

Hunter (1941) assessed the possible role of peripheral representation most definitively in her "fast motor rotation" condition, in which the rat was rotated through 360 degrees between seven and nine times during the retention interval. As Hunter notes, "the speed of the motor was great enough to throw the rat down to one end of the delay box against the door, and to keep the rat there by centrifugal force until the speed slackened [p. 341].'' Although this manipulation surely disrupted any peripheral representations, six of seven rats reached the criterion of 90% accuracy across 30 trials.

Although it is not possible to consider all studies which have addressed the issue of nonmemorial sources of control (see Dember & Fowler, 1958; Petrinovich & Bolles, 1957; Still, 1966), the evidence indicates that performance on the free choice is controlled, at least in part, by a central representation of information derived from the forced run. A corollary of this view is that the decrease in free-choice accuracy evident at longer retention intervals (e.g., 20 and 40 s) reflects (again, at least in part) forgetting of that information.

## THEORETICAL CONCEPTIONS

Conceptualizing the delayed alternation preparation at a theoretical level is our next task. We know that this preparation requires memory. Two issues need to be addressed: (1) the nature of the memorial representation; and (2) the processes which operate upon this representation. Three rather different general frameworks in which to accomplish these joint objectives are discussed below.

### Permanent Memory Conception

Although forgetting occurs over short intervals in delayed alternation, Gordon and Feldman (1978) and Feldman and Gordon (1979) have argued that each time a rat is forced to an arm of the maze, a permanent memorial representation of that experience is established. As Feldman and Gordon state: "According to this view, the permanent storage of information can be viewed as occurring contemporaneously with information input. Thus, all subsequent retrieval would be accomplished from a single, permanent memory store [p. 208]." Because memories are stored permanently, forgetting cannot result from the loss of information from memory storage. Instead, forgetting in delayed alternation is viewed as a product of failure to access or retrieve the appropriate memory as a result of changes in internal or external context which occur during the retention interval. This conception acknowledges no fundamental difference in memory processing between delayed alternation and any other memory preparation regardless of whether the other preparation involves short-term or long-term retention. In essence, Gordon and Feldman are suggesting that the explanatory framework suggested by Spear's (1973, 1978) analysis of long-term retention in the rat is equally applicable to short-term retention.

The parsimony of the position articulated by Gordon and Feldman (1978) is particularly attractive theoretically. Because neither the memorial representation nor memory processing is held to be influenced qualitatively by whether short-term or long-term retention is being assessed, only a single set of explanatory principles is required. Thus, theoretical ideas developed within the context of studies of LTM may be applied directly as a theoretical interpretation of behavior in delayed alternation. This view implies further that highly efficient preparations such as delayed alternation may replace less efficient preparations such as active and passive avoidance in the experimental analysis of animal memory (see Melton, 1963 for a similar argument in the case of human memory).

Although the permanent memory conception is consistent with many findings of delayed alternation research, the viewpoint can be challenged at two levels. At the intuitive level, one might question the notion that the nature of information processing is independent of the sizable procedural differences between delayed alternation and such typical LTM preparations as those involving avoidance training. This point will be elaborated in a subsequent section in which the working memory conception is considered; it is sufficient here merely to suggest the possibility that the nature of information processing may be influenced importantly, and qualitatively, by task requirements.

At the empirical level, one can question whether changes in context are indeed the major source of forgetting in delayed alternation. Numerous studies of long-term retention in the rat have revealed a powerful effect of testing context (e.g., Gordon, 1981; Spear, Smith, Bryan, Gordon, Timmons, & Chiszar, 1980; Welker & McAuley, 1978; Zentall, 1970). In general, these studies suggest that the more closely testing context resembles the context of training the greater is the probability that the training memory controls test performance. In contrast, test performance is independent of the similarity relation between the context of the forcing and the context of the test run in delayed alternation (Grant, 1980). This finding seriously questions the fundamental tenet of the Gordon and Feldman (1978) position, that changes in context are primarily responsible for forgetting in delayed alternation.

## Temporary Memory Conception

The traditional reaction to the criticisms of the permanent-memory conception involves the postulation of multiple memory stores. According to this view, the memory system is in fact two memory systems; a short-term store (STS) in which temporary, labile memories are held and processed, and a long-term store (LTS) in which static, permanent memories reside. Such a general orientation suggests that delayed alternation reflects the processing of temporary memories in STS, whereas performance in LTM preparations reflects the processing of permanent memories retrieved from LTS. In contrast to the Gordon and Feldman (1978) position in which the occurrence of each forced turn establishes a permanent representation and forgetting reflects retrieval failure induced by changes in context, the temporary conception views the occurrence of each forced turn as establishing a temporary representation in STS which fades or decays over time (Roberts & Grant, 1976 have developed a model of pigeon STM within such a general framework, but see Grant, 1981b for a critique of this position). According to this view, forgetting in delayed alternation reflects the loss of information from storage, rather than from reduced accessibility or retrievability.

Inherent in the dual-store conception is the idea that processes different from those which operate on permanent memories in LTS operate on temporary memories in STS. Such a position can easily account for the finding that testing context is a critical determinant of retention in LTM preparations, and is virtually irrelevant in delayed alternation. It is important to note that the distinction between STS and LTS is theoretical, whereas that between STM and LTM is purely operational. Using the terms "STM" and "LTM" does not therefore commit one to a dual-store conception of memory.

Although a number of different models could be developed within this general framework, several convincing critiques of this approach that have appeared recently render these models obsolete. For example, both Lewis (1979) and Spear (1978) have argued that purely temporally based distinctions (as is true of STS and LTS) are neither empirically justifiable nor theoretically useful. Both suggest further that the nature of memory processing is more likely to depend upon the accessibility, rather than the age, of a memory. According to Spear (1978): "the most useful distinction made among aspects of memory processing will probably concern whether the memory is in an 'active' state (and, hence, relatively accessible) or in a 'passive' state (and not accessible without additional information and special processing)" [p. 277].

## Working Memory Conception

Honig (1978) suggests that a working memory conception might be an appropriate framework in which to develop models of short-term retention. His argument is rather straightforward. The position rests upon an acknowledgement of important procedural differences between preparations typically employed to study STM (including delayed alternation) and preparations typically employed to study LTM. Moreover, Honig contends that these procedural differences are causative factors in producing differences in memory processing. Honig's position thus implies that models of delayed alternation must include processes either in addition to or different from those included in models of long-term retention. It is because of the failure of the permanent-memory conception to acknowledge such differences in processing that that view is not favored here.

To elaborate on the procedural differences between preparations typically employed to study STM and those employed to study LTM, consider first LTM preparations in which the reinforcement contingencies remain relatively stable from one occasion to the next. Such contingencies might specify, for example, that a right turn at the choice point leads to food in a spatial discrimination, or that crossing to the other compartment within 10 s of tone onset precludes shock in an avoidance task. Honig (1978) has suggested that once the appropriate associative structure(s) has been formed, subsequent performance requires only that the appropriate reference memory be accessed or retrieved.

In delayed alternation, on the other hand, the reinforcement contingencies are more dynamic in that the response which is reinforced at testing (i.e., turning right or left on the free-choice run) is dependent critically upon the nature of a cue presented earlier (i.e., the direction of the forced turn). Thus, in addition to reference memory (which presumably codes the contingencies "right forcing—left free" and "left forcing—right free"), delayed alternation also requires that information derived from the forced turn be maintained over short intervals and be updated between trials. Honig (1978) has used the term "working memory" to describe these additional requirements of delayed conditional discrimination tasks. According to the working-memory view, the dynamics of the reinforcement contingencies in delayed alternation invoke a similar dynamism in memory processing. Thus, processes not required in simple (nonconditional) discriminations are likely operative in conditional discriminations such as delayed alternation.

## A MODEL OF SHORT-TERM MEMORY IN RATS

The model of STM considered here is based on delayed alternation data and was guided by the working memory point of view. The model is neither a final theoretical statement nor a particularly original formulation. It is perhaps best viewed as a coalescence of a number of working assumptions and hypotheses concerning rat STM. As such, its organizing capabilities and heuristic value will likely exceed its explanatory power. After the nature of the memorial representation is considered, there is a discussion of the processes held to operate on that representation.

### The Nature of the Memorial Representation

The model holds that two multidimensional representations are established early in delayed alternation training. The two representations are distinguished on the basis of detailed information derived specifically from the forced turn. The representations are viewed as permanent in that the specific attributes that constitute the representations are not lost from storage. Although permanently represented, the attributes are held to move bidirectionally between two discrete states of activation: an active state and an inactive (or passive) state.

*Memories as Multidimensional.*    In 1978, Spear remarked that "it was first realistic, then useful, and now customary to take a multidimensional view of a memory [p. 4]." Presently, the multidimensional view is perhaps more obligatory than customary. The multidimensional view of memory has proven exceedingly useful in the theoretical analysis of such long-term retention phenomena as warmup decrement and its alleviation, reminder effects, context-cuing, amnesia, and hypermnesia (see Spear, 1978). In addition, the usefulness of this view of memory will become apparent in the course of the description and evaluation of the model of delayed alternation being developed here.

The multidimensional view implies that a memory is composed of a collection of attributes which represent all aspects of an episode that are noticed by an organism. Thus, a memory is a collection of associatively related components, or attributes, that each represent "a psychologically separable event" (Spear, 1978, p. 4) comprising an episode. To illustrate this view, it may be useful to cite an example employed by Spear:

> A rat confronted with the task of leaving one compartment and entering another within five seconds to avoid footshock may represent this episode collectively in terms of separate memory attributes (apart from the immediate experimental contingencies governing the footshock): the odor of the room, the sensation of the experimenter's hand, the rat's internal hormonal state based upon time of day or point in estrous cycle or state of adrenal depletion, the rat's physical or emotional state (illness, fatigue, hunger, or fear), the severity of the footshock, the structure of the grid floor, and perhaps the sequence of just-preceding events in its home cage leading up to its predicament [p. 4].

Given that the representations in delayed alternation are multidimensional, it is appropriate to inquire as to the number and type of attributes which constitute the memories. I will focus on the nature of the attribute (or attributes) which distinguish the two permanent representations. These attributes must in some way represent information derived from the forced run, and may be referred to as "target" attributes, in that activation of these attributes is critical to accurate test performance. It should be clear that in addition to the target attribute(s) which distinguish the two permanent representations, each representation will also contain a number of common nontarget, or contextual, attributes. For example, each of the two representations will likely contain attributes representing information concerning the reinforcer (amount, palatability, texture, or other qualities), brightness of the maze, odors present in the testing room, texture of the maze floor, and general illumination levels. Because these contextual attributes are common to each of the two representations, they need not be considered further here.

Unfortunately, virtually nothing is known about the specific form in which information derived from the forced run is coded. One issue is whether to-be-remembered information is coded prospectively or retrospectively, an issue prominent in recent studies of pigeon STM (e.g., Grant, 1981b, 1982a; Honig, 1978; Roitblat, 1980, 1982). That is, does the rat remember what happened (e.g., "a right turn occurred on the forcing"), or what to do at testing (e.g., "turn left on the free choice"), or both? A second issue concerns the dimension(s) involved in the target attribute(s). For example, does the code involve information about places either visited on the forcing or to-be-visited on the free choice, or does it involve information about response direction, or are both types of information represented?

Although experimental analysis of these issues would require modification to the typical delayed alternation preparation, the answers obtained would permit inferences about coding in alternation. It must be emphasized that such an experimental program is envisaged as determining the flexibility of coding, rather than its inevitable form. That is, it is unlikely that a single form of coding is employed in all situations. Rather, the nature of coding is likely importantly influenced by the characteristics of the task (see Restle, 1957). If so, interest should focus on identifying variables which influence coding and on specifying the nature of the resultant codes.

Given the paucity of studies addressing the issue of coding processes in delayed alternation, the present model cannot describe the specific nature of target attributes. It is important to note, however, that the present conception endorses two general propositions concerning coding. First, it is held that each of the permanent representations may contain more than one target attribute. Second, given that a permanent representation is comprised of two or more target attributes, each target attribute is held to be activated (i.e., retrieved) and inactivated (i.e., forgotten) independently. This view implies that rate of forgetting is determined in part by the number of available target attributes.

Although it is not feasible to speculate on the nature of the target attributes at present, subsequent discussion requires a means of differentiating between the two permanent representations. To accomplish this, the permanent representation established during training as a result of forcings to the right will be referred to as the "A" memory. Similarly, the permanent representation established during training as a result of forcings to the left will be referred to as the "B" memory. Adopting these arbitrary symbols will permit us to remain neutral concerning the number and nature of target attributes.

*Active and Inactive Memories.*    The two permanent representations (or, more appropriately, the attributes comprising those representations) are held to exist in one of two discrete states of activation: active and inactive (for similar views see Gordon, 1981; Grant, 1981b; Lewis, 1979; Spear, 1978). Active memories are those currently receiving processing and are thus highly accessible. Inactive memories are those not receiving processing and are thus less accessible. Memories are held to move bidirectionally between these two states of activation during the course of training and testing. Only currently active memories can exert control over behavior.

## Memory Processes

In this section, five processes of memory of prime importance in delayed alternation are considered. The first process, involving memory retrieval, operates on inactive memories only, whereas the other four processes—rehearsal, tagging, updating, and discrimination—influence only active memories. Memory retrieval results in the activation of an inactive attribute, and maintenance rehearsal controls the duration of residence in the active state. Memory tagging results in attributes being appended temporarily to an active representation. Memory updating involves the reprocessing of a currently active representation such that new tags are appended and rehearsal is reinitiated. Finally, on those occasions in which conflicting target attributes are active at testing, processes of discrimination determine which attribute controls performance. Description of these processes will complete specification of the model.

*Retrieval.*    The two permanent representations (A coding a right turn and B coding a left turn) reside normally in an inactive state and are provoked to a state of temporary activation by processes of retrieval. The occurrence of a run to an arm of the maze, whether forced or free, is viewed as a necessary, although not inevitably sufficient, condition for target attribute activation. Specifically, the probability of an inactive target attribute moving to an active state is greater than zero only during (or immediately after) a run to an arm of the maze. The occurrence of such a run results in activation of a target attribute associated with a turn in that direction (right-A or left-B) with some probability less than unity and may, on rare occasions, result in activation of a target attribute comprising the alternate permanent representation (right-B or left-A). The activation of a particular target attribute is held to be partially independent of the activation of other target attributes which may comprise the permanent representation. On any particular trial, then, only a portion of the target attributes associated with a permanent representation may have been activated.

The present model departs from that of Gordon and Feldman (1978) in suggesting that forced runs do not establish a permanent representation during memory testing. Gordon and Feldman suggest that a forcing to the right or left necessarily establishes a permanent representation of that episode. On the other hand, I am suggesting that once the A and B representations have been established (a process which presumably occurs early in training), a forced turn activates one of the two permanent representations; it does not establish its own permanent representation.

Because I have previously suggested the possibility that free-choice performance may be controlled by memories retrieved at testing when sufficiently long retention intervals are employed (Grant, 1980), it is important to emphasize that the present model views retrieval as occurring only as a function of a run to an arm of the maze. Thus, any attribute not active at the end of the retention interval will remain inactive at the time the free-choice response is executed. Test performance can therefore be controlled only by target attributes which have remained active during the retention interval.

There are two reasons for assuming that neither the A nor B representation is activated at testing. The first is parsimony. The model being developed here is viewed as a complete account of delayed alternation behavior, and thus incorporating the notion that memories can be retrieved at testing would not enhance explanatory power. Second, as discussed in more detail later, data obtained by Gordon, Taylor, and Mowrer (1981, Experiment 3) may be viewed as problematic for the notion that memories are retrieved at testing.

*Maintenance Rehearsal.*    Maintenance rehearsal is viewed here as a flexible, controlled process which prolongs the time period during which an active memory remains in that state (see Grant, 1981b, 1984 and Maki, 1981 for further discussion of rehearsal in animals). The present model views rehearsal in delayed alternation as occurring centrally and tending to decrease in probability and/or intensity as a function of time since initiation. The tendency for rehearsal to wane over time is not viewed as occurring inevitably, and this trend may be reversed under conditions described in the section on memory updating.

The suggestion that rehearsal is a controlled process implies that contingencies of reinforcement determine in part the conditions under which rehearsal is initiated. In the case of delayed alternation, activation of the A or B representation by a forced turn is likely to be followed by rehearsal because maintaining that representation in an active state enhances the probability of reinforcement on the test run. Similarly, it may be anticipated that the occurrence of a free-choice run would terminate any ongoing rehearsal and, moreover, would fail to provoke rehearsal if either the A or B representation was activated by the free-choice run.

As developed to this point, the model suggests that after a period of training, a forced turn will likely activate the appropriate permanent representation (a right forcing activating the A representation and a left forcing activating the B representation). If the appropriate representation is activated at the time of the forcing, a correct alternation will occur at testing if that representation (or, more accurately, one or more of the target attributes associated with that representation) has remained in an active state throughout the retention interval. In determining whether that representation remains active, the probability of maintenance rehearsal initiation is important and, if initiated, the duration and/or intensity of rehearsal.

*Tagging.* At the time of target attribute activation, one or more additional attributes may be appended temporarily to the active representation; a process referred to as tagging. In contrast to target attributes, these additional attributes, called memory tags, are not represented permanently and therefore cannot be retrieved. Memory tags code information which is to some extent unique to a particular occurrence of a to-be-remembered episode. One type of tag postulated commonly is a time tag which is held to represent the temporal properties associated with the occurrence of a to-be-remembered episode.

Although the model holds that temporal attributes are always represented as tags in delayed alternation, some types of information may be represented either temporarily as tags or permanently as target attributes. Whether a characteristic is represented permanently or is appended as a tag depends upon the stability of that characteristic across occurrences of a to-be-remembered episode. Invariant properties of a to-be-remembered episode are represented permanently, whereas more labile properties are represented as tags. Suppose that for some animals the odor of almond always accompanied a forcing to the right and the odor of rose always accompanied a forcing to the left. In another group, suppose that only one of the two odors was presented on each run, but that the odor presented and the direction of the run were uncorrelated. In the former case the odors would be represented permanently, and in the latter case would be represented temporarily as tags.

On some trials, then, one or more attributes may be appended temporarily to an active representation. On such occasions, the rehearsal status of the tag is dependent entirely on the rehearsal status of the target attribute which is currently active. A tag is rehearsed only if the currently active target attribute is rehearsed. However, the ultimate fate of target attributes and tags is held to be different. Target attributes return ultimately to the inactive state, and forgetting of such attributes is therefore a function of a change in activational state (i.e., from active to inactive). On the other hand, tags do not move to an inactive state, and are instead either active or nonexistent (they disappear without leaving a trace). Forgetting of information represented as a tag is thus viewed as resulting from memory loss.

The suggestion that information represented as a tag is eventually lost from storage may at first appear incompatible with the notion that tags are often rehearsed. Notice, however, that the sole function of maintenance rehearsal is to increase the period of time during which an active memory remains in that state; maintenance rehearsal does not in any way promote the formation of permanent memories. The formation of permanent memories is held to be influenced by a qualitatively different form of rehearsal that, for lack of a better term, is referred to as associative rehearsal. It is this latter type of rehearsal which is responsible, as an example, for the permanent representations established early in delayed alternation training. Although a distinction between qualitatively different types of rehearsal has not been prominent in treatments of animal memory (but see Grant, 1984 and Maki, 1979), it is suggested here that this is only temporary and that such a distinction may prove to be of central importance to our understanding of memory processing.

Thus, another fundamental contrast between the present model and that of Gordon and Feldman (1978) concerns the nature of the processing evoked by the occurrence of a forced turn. The present view is that this event triggers a type of processing, maintenance rehearsal, fundamentally different from that evoked in associative learning tasks (a process referred to as associative rehearsal). Gordon and Feldman hold that the occurrence of a forced turn establishes a permanent representation on each occasion, a view which implies that the occurrence of a forced turn evokes the same type of processing as that evoked by any other episode.

*Updating.*    Memory updating involves three component processes which can only be executed sequentially, although the sequence may be terminated at any point. These component processes are referred to as context comparison, retagging, and maintenance rehearsal reinitiation. The easiest way to describe these processes is through comparison with those of retrieval, tagging, and maintenance rehearsal initiation. It has been argued previously that certain environmental events may activate one or more target attributes, and that this activation may be followed closely by tagging and the initiation of rehearsal. In the case of memory updating, a subsequent event on a trial results in the target attribute which was activated earlier being retagged and rehearsal being re-initiated. Memory updating does not, therefore, involve retrieval, because the requisite target attribute must already be in an active state. Updating involves instead comparison of contemporary context with that represented by the tags appended to the currently active target attribute(s). If the contexts are discriminably different, then new tags are appended and rehearsal is reinitiated, thus completing the memory-updating sequence.

Two events which trigger the initial component process in memory updating—context comparison—may be identified at present. The first is that a run to an arm of the maze occur in the presence of an active target attribute associated with that run. The second is that goal-box placement occur in the presence of an active target attribute associated with a run to that goal box (or arm). In both cases, retagging and rehearsal reinitiation will occur only if the contemporary context of the run or goal-box placement differs discriminably from that which was present when the currently active target attribute was retrieved. Thus, a run to an arm of the maze may result in memory retrieval or updating, depending upon whether the associated target attribute is inactive or active when the run occurred. On the other hand, goal-box placement can never provoke attribute activation (see the section on retrieval), but may initiate memory updating if a target attribute associated with a run to that goal box is currently active.

Processes of memory updating may thus be viewed as refreshing or rejuvenating an active representation. At least two mechanisms would be expected to lead to enhanced retention of an updated target attribute. First, updating the time tag should facilitate discrimination between that target attribute and any conflicting target attributes which may have remained from prior trials (see the section on discrimination). Second, reinitiation of rehearsal should restore rehearsal to its initial level of intensity, thus extending the period during which the target attribute is maintained in the active state.

*Discrimination.* It was suggested that only active attributes can control performance, and that attributes are not activated at testing. Assuming that target attributes comprising the appropriate representation only are retrieved or activated by the forcing, the free-choice response will be correct if at least one of those attributes has remained active until testing. On the other hand, free-choice accuracy will be at chance if no target attributes are active at testing. Suppose, however, that one or more target attributes belonging to each of the permanent representations is active at testing. Such a situation would arise if attributes activated on earlier trials remained in an active state at the time of testing on the current trial, and could be induced experimentally by administering a forcing to each arm prior to the test. It is in cases in which conflicting target attributes are active at testing that processes of discrimination are operative. Discrimination simply refers to the processes whereby one active target attribute comes to control choice responding in deference to a second, antagonistic active target attribute.

In some cases, discrimination may be based on properties of the target attributes themselves. Suppose a particular rat has a left-turn preference and has active target attributes representing "go left" and "go right." The preference may result in the "go-left" attribute controlling performance. In the more typical case, and the more interesting case from the perspective of memory processing, the discrimination is shaped by the contingencies of reinforcement. In such cases, discrimination is based upon information represented by the tags associated with the target attributes. Consider that the contingencies of delayed alternation specify a relation between the direction of the most recent forcing and the direction of the free-choice response which procures a reinforcer. The contingencies are also such that the direction of turns other than the most recent are irrelevant to obtaining a reinforcer on the current free-choice run. It is therefore suggested that the discrimination process involves "reading" time tags and the control of performance by the target attribute associated with the most recent time tag.

## EVALUATION OF THE MODEL

The model is evaluated by considering its explanatory power, heuristic value, and generality. The following discussion does not treat these aspects of the model exhaustively, but rather in an illustrative manner.

## Explanatory Power

If the model is to enjoy even a modicum of success, then it must have some explanatory power. This aspect of the model is considered through a brief theoretical treatment of phenomena in delayed alternation. The opportunity has arisen previously to apply the model as an interpretation of context effects (Grant, 1980), proactive interference, (Grant, 1981a), and the influence of forget and remember cuing (Grant, 1982b). The present discussion is therefore restricted to phenomena not considered previously.

Roberts (1972b) found that rate of forgetting was reduced by rendering the arms of a Y-maze visually distinct. Rats in the visual cues group encountered a black card at the entrance to one goal box and a white card at the entrance to the other goal box (the cards remained in the same positions throughout training and testing). In the no-visual cues group, cards of the same brightness were present at each entrance. According to the present view, the permanent representations formed in the visual-cues group contained an additional target attribute based upon information derived from the visual distinctiveness of the goal-box entrances. Because attributes are held to be activated and inactivated with some degree of independence, a reduced rate of forgetting would derive from those occasions on which only the "visual" target attribute was active at testing.

In a subsequent series of experiments employing a T-maze, Roberts (1974) assessed the effect of administering the forced turn once, twice, or four times prior to the retention test. It was found that free-choice accuracy was enhanced equivalently at retention intervals of 0, 45, and 120 s by repetition of the forced turn. Because the occurrence of any particular forced turn is held to activate an associated target attribute with a probability less than unity, the beneficial effect of repetition is interpretable readily as a product of an increased probability of target attribute activation prior to onset of the retention interval.

More intriguing was Roberts' (1974) finding that spaced repetition (60 s between successive forcings) resulted in less rapid forgetting than did massed repetition (successive forcings followed in immediate succession). According to the present model, this finding derives from a higher probability of memory updating when repetitions are spaced rather than massed. Under conditions of massed repetition, updating is unlikely because there is little contextual variability within the sequence of forcings. On the other hand, introducing an interval between forcings would increase contextual dissimilarity and thereby increase the probability of memory updating. This discussion may be rendered less abstract by considering only the temporal aspect of context. It is then clear that a rat is more likely to discriminate between the time-at-activation and the time-at-present-forcing under conditions of spaced repetition.

The memory-updating interpretation anticipates correctly that the beneficial effects of spacing should be manifested as a decrease in rate of forgetting.

This is the case because the reinitiation of rehearsal and appending a new time tag should each reduce the rate of forgetting. Also confirmed by Roberts' (1974) data is the prediction that spacing should fail to enhance retention if the spacing interval is too short to permit discriminable changes in context from occurring between repetitions. In support of this view, Roberts found no difference in retention as a function of massed versus spaced repetition when a spacing interval of only 20 s was employed. On the other hand, spacing repetitions either 60 or 180 s apart resulted in better retention than did massing repetitions.

The notion of memory updating may be applied as well as an interpretation of reminder treatment effects. Gordon and Feldman (1978) and Feldman and Gordon (1979) have shown that retention of information derived from a forcing is enhanced by placement in that goal box at some point during the retention interval (see also Grant & Marshal, 1985). The present view of this phenomenon is not that the reminder treatment activated inactive attributes, but rather that the treatment initiated processes of memory updating on those occasions on which at least one target attribute remained active at the time of goal-box placement.

In a recent study, Gordon, Taylor, and Mowrer (1981, Experiment 3) found that although a 5-s exposure to the goal box to which the animal had been previously forced served as an effective reminder, a similar exposure to the start box and runway did not. This finding is consistent with the present view in that start box and/or runway exposure is held to provoke neither target attribute activation (retrieval) nor memory updating. Perhaps more importantly, this finding is rather damaging to the view that retention-test performance is controlled by memories retrieved at the time of testing. How can one reconcile the finding that testing context is not an effective retrieval cue with the view that free-choice performance is typically controlled by target attributes retrieved at testing? In the present view, such a reconciliation is neither feasible nor necessary.

## Heuristic Value

Although several testable predictions may be derived from the model, only a few examples are discussed here. Consider first the influence of context on free-choice performance. According to the present view, aspects of context are appended temporarily as tags and therefore serve only to increase the discriminability between active memories. Free-choice performance should thus be independent of the degree of contextual similarity between a forcing and the test. On the other hand, the views of Roberts (1974) and of Gordon and Feldman (1978) suggest that context is incorporated into a representation which may be retrieved at testing. If it is assumed that performance is controlled by attributes retrieved at testing only in the absence of an active target attribute, then testing context should exert progressively stronger control over free-choice accuracy as the retention interval is lengthened. Although no evidence of such an effect was found when retention intervals of 0 and 20 s were employed (Grant, 1980), a test in which longer intervals are employed is required. According to the present view, testing context should not influence performance at any retention interval length.

A second way in which to test whether attributes are sometimes retrieved at testing involves using the Gordon et al (1981) technique of interpolating an exposure to the start box and runway of the maze during the retention interval. Although they found this treatment to be ineffective as a reminder, the forcing/cuing interval was relatively short (45 s). If, contrary to the present position, attributes are retrieved at testing when long retention intervals are employed, then such a treatment should prove to be an effective reminder at longer forcing/treatment intervals.

As a second example of the heuristic value of the model, consider the present view of the effects of spacing repetitions of a forced turn. As pointed out earlier, the memory-updating interpretation correctly anticipates that spacing will enhance retention only when the spacing interval is sufficiently long to permit discriminable changes in context to occur between repetitions of the forcing. The present view predicts further, however, that increases in the spacing interval beyond that range of values will reverse the difference between massed and spaced repetition. In other words, if the spacing interval is long enough to ensure that a target attribute activated on an earlier forcing is never active at the time of a subsequent forcing, then massed repetition should result in higher accuracy than spaced repetition. This is the case because: (1) memory updating will never occur on spaced repetition trials; and (2) the probability of target attribute activation at the end of the forcing sequence will be higher on massed repetition trials.

As a final example, the present interpretation predicts that the beneficial effect of a reminder treatment will be enhanced by increasing the contextual dissimilarity (within limits) between the original forcing and the goal-box placement. This is the case because retagging and the reinitiation of rehearsal are held to occur only if the forcing and reminder are administered in discriminably different contexts. A reminder treatment administered shortly after the forcing should also have no effect. Such a reminder could be rendered effective, however, by altering the context between the forcing and the goal-box placement. This could be accomplished, for example, by manipulating background auditory or olfactory cues.

## Generality

The question of generality can be addressed both within-species and between-species. For within-species, the model may prove of value in the theoretical analysis of data obtained in preparations other than delayed alternation. Although other preparations have been employed in the experimental analysis of rat STM (e.g., Wallace, Steinert, Scobie, & Spear, 1980), the data base is far too limited to permit an argument for within-species generality to be supported empirically. It can only be suggested that, although rats may process information with differential efficiency across modalities, it is unlikely that fundamentally dissimilar systems have evolved to process information from different modalities.

It is anticipated also that the model will prove general across species, although the argument is no less speculative than in the case of within-species generality. In many respects the present model is highly similar to a model of pigeon STM advanced recently by the author (Grant, 1981b). For example, both models were formulated within Honig's (1978) working memory conceptual framework, suggest that to-be-remembered events temporarily activate permanent representations, and emphasize processes of maintenance rehearsal. Whether this resemblance is more a product of a limited understanding of STM than it is an indication of a fundamental continuity in information processing systems across species is an issue yet to be resolved.

It may be argued, however, that any attempt to formulate a model applicable across species is rendered futile by the demonstration of between-species differences in functional relations. As an example of such a functional dissimilarity, consider the influence of spacing repetitions on retention in rats and pigeons. In contrast to rats, pigeons demonstrate enhanced performance when repetitions of a to-be-remembered event are massed rather than spaced (Roberts, 1972a; Roberts & Grant, 1974). One possible theoretical response would be to suggest that processes of memory updating operate in rats but not in birds. This strategy would require abandoning the notion that similar processes operate in the two species, and thus is not favored here.

An alternative theoretical response involves retaining the notion of similar processes, and attributing the differential influence of spacing to differences in experimental procedures and/or model parameters. The specific suggestion is that both rats and pigeons are capable of updating memory, but that delayed alternation is more likely to invoke those processes than is delayed matching. One reason for this might be differences in contextual variability; testing in a T-maze involves greater between-repetition variability than does testing in the highly controlled environment of an enclosed chamber. If so, context comparison would lead to a higher probability of retagging and rehearsal reinitiation in rats.

A second factor, perhaps acting synergistically with the first, that could produce a difference in the probability of memory updating involves possible differences in the probability of a target attribute remaining active between repetitions. It is well documented that rate of forgetting is markedly greater in pigeons tested in delayed matching than it is in rats tested in delayed alternation (perhaps reflecting differences in the efficiency of maintenance rehearsal). It thus follows that in pigeons spacing intervals sufficiently long to permit discriminable contextual variability between repetitions may at the same time be sufficiently long to permit a previously activated target attribute to return to the inactive state. If so, the operation of processes of memory updating would be precluded.

My point has not been to provide a definitive treatment of spacing effects in rats and pigeons. The intent was rather to illustrate one approach in dealing theoretically with between-species differences in empirical relations within a general model of STM. Whether this approach will prove successful remains an open question.

## ACKNOWLEDGMENT

Preparation of this chapter was supported by Grant A0443 from the Natural Sciences and Engineering Research Council of Canada.

## REFERENCES

D'Amato, M. R. Delayed matching and short-term memory in monkeys. In G. H. Bower (Ed.), *The psychology of learning and motivation, Vol. 7.* New York: Academic Press, 1973.

Dember, W. N., & Fowler, H. Spontaneous alternation behavior. *Psychological Bulletin,* 1958, *55,* 412–428.

Estes, W. K., & Schoeffler, M. S. Analysis of variables influencing alternation after forced trials. *Journal of Comparative and Physiological Psychology,* 1955, *48,* 357–362.

Feldman, D. T., & Gordon, W. C. The alleviation of short-term retention decrements with reactivation. *Learning and Motivation,* 1979, *10,* 198–210.

Gordon, W. C. Mechanisms of cue-induced retention enhancement. In N. E. Spear & R. R. Miller (Eds.), *Information processing in animals: Memory mechanisms.* Hillsdale, NJ: Lawrence Erlbaum Associates, 1981.

Gordon, W. C., & Feldman, D. T. Reactivation-induced interference in a short-term retention paradigm. *Learning and Motivation,* 1978, *9,* 164–178.

Gordon, W. C., Taylor, J. R., & Mowrer, R. R. Enhancement of short-term retention in rats with pretest cues: Effects of the training-cueing interval and the specific cueing treatment. *American Journal of Psychology,* 1981, *94,* 309–322.

Grant, D. S. Delayed alternation in the rat: Effect of contextual stimuli on proactive interference. *Learning and Motivation,* 1980, *11,* 339–354.

Grant, D. S. Intertrial interference in rat short-term memory. *Journal of Experimental Psychology: Animal Behavior Processes,* 1981, *7,* 217–227. (a)

Grant, D. S. Short-term memory in the pigeon. In N. E. Spear & R. R. Miller (Eds.), *Information processing in animals: Memory mechanisms.* Hillsdale, NJ: Lawrence Erlbaum Associates, 1981. (b)

Grant, D. S. Prospective versus retrospective coding of samples of stimuli, responses, and reinforcers in delayed matching with pigeons. *Learning and Motivation,* 1982, *13,* 265–280. (a)

Grant, D. S. Stimulus control of information processing in rat short-term memory. *Journal of Experimental Psychology: Animal Behavior Processes,* 1982, *8,* 154–164. (b)

Grant, D. S. Rehearsal in pigeon short-term memory. In H. L. Roitblat, T. G. Bever, & H. S. Terrace (Eds.), *Animal cognition.* Hillsdale, NJ: Lawrence Erlbaum Associates, 1984.

Grant, D. S. & Marshal, M. L. Short-term retention in rats: The effect of goal-arm confinement on delayed alternation performance. *Animal Learning & Behavior*, 1985, *13*, 109–115.

Honig, W. K. Studies of working memory in the pigeon. In S. H. Hulse, H. Fowler, & W. K. Honig (Eds.), *Cognitive processes in animal behavior*. Hillsdale, NJ: Lawrence Erlbaum Associates, 1978.

Hunter, R. R. Symbolic performance of rats in a delayed alternation problem. *Journal of Genetic Psychology*, 1941, *59*, 331–357.

Lewis, D. J. Psychobiology of active and inactive memory. *Psychological Bulletin*, 1979, *86*, 1054–1083.

Maki, W. S. Discrimination learning without short-term memory: Dissociation of memory processes in pigeons. *Science*, 1979, *204*, 83–85.

Maki, W. S. Directed forgetting in animals. In N. E. Spear & R. R. Miller (Eds.), *Information processing in animals: Memory mechanisms*. Hillsdale, NJ: Lawrence Erlbaum Associates, 1981.

Melton, A. W. Implications of short-term memory for a general theory of memory. *Journal of Verbal Learning and Verbal Behavior*, 1963, *2*, 1–21.

Petrinovich, L., & Bolles, R. Delayed alternation: Evidence for symbolic processes in the rat. *Journal of Comparative and Physiological Psychology*, 1957, *50*, 363–365.

Restle, F. Discrimination of cues in mazes: A resolution of the "Place-versus-response" question. *Psychological Review*, 1957, *64*, 217–228.

Roberts, W. A. Short-term memory in the pigeon: Effects of repetition and spacing. *Journal of Experimental Psychology*, 1972, *94*, 74–83. (a)

Roberts, W. A. Spatial separation and visual differentiation of cues as factors influencing short-term memory in the rat. *Journal of Comparative and Physiological Psychology*, 1972, *78*, 284–291. (b)

Roberts, W. A. Spaced repetition facilitates short-term retention in the rat. *Journal of Comparative and Physiological Psychology*, 1974, *86*, 164 -171.

Roberts, W. A., & Grant, D. S. Short-term memory in the pigeon with presentation time precisely controlled. *Learning and Motivation*, 1974, *5*, 393–408.

Roberts, W. A., & Grant, D. S. Studies of short-term memory in the pigeon using the delayed matching to sample procedure. In D. L. Medin, W. A. Roberts, & R. T. Davis (Eds.), *Processes of animal memory*. Hillsdale, NJ: Lawrence Erlbaum Associates, 1976.

Roitblat, H. L. Codes and coding processes in pigeon short-term memory. *Animal Learning and Behavior*, 1980, *8*, 341–351.

Roitblat, H. L. The meaning of representation in animal memory. *The Behavioral and Brain Sciences*, 1982, *5*, 353–406.

Spear, N. E. Retrieval of memory in animals. *Psychological Review*, 1973, *80*, 163–194.

Spear, N. E. *The processing of memories: Forgetting and retention*. Hillsdale, NJ: Lawrence Erlbaum Associates, 1978.

Spear, N. E., Smith, G. J., Bryan, R. G., Gordon, W. C., Timmons, R., & Chiszar, D. A. Contextual influences on the interaction between conflicting memories in the rat. *Animal Learning and Behavior*, 1980, *8*, 273–281.

Still, A. W. Repetition and alternation in rats. *Quarterly Journal of Experimental Psychology*, 1966, *18*, 103–108.

Wallace, J., Steinert, P. A., Scobie, S. R., & Spear, N. E. Stimulus modality and short-term memory in rats. *Animal Learning and Behavior*, 1980, *8*, 10 -16.

Welker, R. L., & McAuley, K. Reductions in resistance to extinction and spontaneous recovery as a function of changes in transportational and contextual stimuli. *Animal Learning and Behavior*, 1978, *6*, 451–457.

Zentall, T. R. Effects of context change on forgetting in rats. *Journal of Experimental Psychology*, 1970, *86*, 440–448.

# 9  Memory Theories: Past, Present, and Projected

Douglas L. Medin
Gerald I. Dewey
*University of Illinois*

This chapter does not present a new theory of animal memory, nor does it produce a review and analysis of extant theories. Instead, the meta-theoretical assumptions associated with research on animal memory, how these assumptions have changed over the last few decades, and what motivated these changes is discussed in general terms.

A rough analogy may convey our intentions. Assume that advances in understanding animal memory can be measured in terms of distance traveled by a boat along a stream. To someone who has just boarded this boat, progress may be difficult to gauge. By looking at where the boat has been, one may obtain a greater appreciation of the advances that have been made. The future is difficult to discern and it may not be worthwhile to venture precise projections. However, our aim in attempting to look ahead is to anticipate possible obstacles that might be avoided by careful steering.

## OVERVIEW

The chapter is divided into three main sections corresponding to past perspectives on animal memory, present conceptions, and an outline of some prospects for future directions. The first section (the past) is concerned with what may be termed the study of memory from a learning perspective. We consider assumptions associated with this approach and then detail some of the research that led investigators to abandon this relatively narrow view of memory. The second section (the present) attempts to characterize several contemporary views of research on animal memory. The current era easily could be described as learning studied from a memory perspective. In this section, we provide a review of current activity leading up to what we view as three potential problems or limitations that, together, constitute an agenda for future research. These three topics are described in more detail in the third main section of this paper.

## THE PAST—MEMORY AS ASSOCIATIVE STRENGTH

Learning and memory are closely intertwined concepts. By definition, it seems true that without learning, new memories could not be developed and without memory, old learning could not be manifest. Until recently, interactions between learning research and memory research were very asymmetrical, consisting mainly of the use of learning concepts to interpret memory phenomena. Constructs such as response competition, spontaneous recovery, and proactive and retroactive inhibition were lifted from one domain (learning) and applied to the other.

Associated with this use of learning constructs in memory research was a definition of memory that was decidedly biased toward a learning orientation: Memory was defined primarily as the performance of a learned act. Ruggierio and Flag (1976) refer to this as stimulus–response memory, where memory is measured in terms of the strength of the stimulus–response association. In the remainder of this section, four assumptions associated with this approach to memory are developed and then evidence bearing on (and undermining) these assumptions is reviewed.

### Assumptions

Our particular list of assumptions associated with stimulus–response memory is not the only possible list, and other reviewers might suggest more or fewer assumptions or state them somewhat differently. However, the list does seem to convey the flavor of work completed in this genre, and evidence relevant to the present list would be equally relevant to alternative statements of the learning approach to memory.

The first assumption was that learning is basically a matter of forming stimulus–response associations. Although considerable debate took place concerning whether or not reinforcement was necessary for learning, there was general agreement that stimulus traces came to be associated with particular responses. Much of the early work on delay of reinforcement was motivated by the idea that, if learning was to occur, reward had to come before the stimulus trace and response trace faded (often, the optimal interval turned out to be less than a second). A minority opinion was that stimulus-stimulus associations could be formed (see Bolles, 1975, 1976, for reviews) but even these views endorsed the idea that stimulus traces entered into associations.

A second, closely related assumption was that memory could be conceived of as the performance of a learned act. As John B. Watson put it, "By 'memory,' then, we mean nothing except the fact that when we meet a stimulus again after an absence, we do the old historical thing . . . that we learned to do when we were in the presence of that stimulus in the first place" (Watson, 1924, p. 237).

The third assumption was that performance is a function of the strength of associations. When multiple responses are associated with a single stimulus, then the response with the greatest associative strength is most likely to occur in response to that stimulus.

The last assumption was that forgetting can be attributed either to the interference with (e.g., response competition) or decay of associative strength. Again, debate centered around decay versus interference but virtually everyone agreed that trace strength was the changing commodity.

## Problems with the Learning Approach to Memory

So as not to be misunderstood, we probably should state explicitly our opinion that memory research has yielded, is yielding, and will continue to yield cumulative knowledge, even as theoretical orientations develop, shift, or are dropped. For example, interference theory was developed in the learning tradition of the 1930s and 1940s and it continues to appear in various guises and to be influential in contemporary theorizing (e.g., Anderson, 1976, 1981; Anderson & Bower, 1973). This general faith in progress aside, our main purpose is to describe changes in theoretical perspectives, and (in the following paragraphs) findings that reveal severe limitations of each of the assumptions associated with the learning approach to memory are reviewed.

### Stimulus Traces Versus Memory Traces

The idea that stimulus traces enter into associations was consistent with early observations that learning did not occur unless reinforcement followed very quickly. Now, however, there is growing evidence that learning can take place over quite long delays. Initially, evidence on long delay learning was drawn from experiments on taste aversion learning and thought (by some) to reflect a highly specialized mechanism, but now it is clear that long-delay learning is a quite general phenomenon (see D'Amato, Safarjan & Salmon, 1981; Garcia, Ervin, & Koelling, 1966; Revusky, 1971, for relevant literature). The observation that learning may take place over long delay intervals suggests that instead of confining associations to stimulus traces, one should allow for associations to be formed to memory traces (see Honig, 1978, for related discussion).

The idea that memory traces enter into associations has been incorporated by Capaldi into his insightful analyses of sequential behavior (e.g., Capaldi, 1971). Rats are presented with a series of trials involving runs down an alley that may result in reward or nonreward. Capaldi has shown that an excellent account of a variety of observations such as speed of responding and rate of extinction is provided by the idea that the memory for a preceding trial outcome may be associated with events on the current trial. For example, when reward and nonreward alternate, rats come to run fast when they are about to receive a reward and run slowly when they will not. According to Capaldi's theory, the memory for nonreward on a preceding trial paired with reward on the current trial leads to a conditioned anticipation of reward following a nonrewarded trial.

Physiologically-oriented research is also benefitting from the view that memory traces enter into learning. Early work on retrograde amnesia discovered fairly sharp "delay of disruption gradients" for amnesic agents such as electroconvulsive shock (ECS), somewhat analogous to delay of reinforcement gradients. But Lewis and his associates (e.g., Lewis, 1976; Lewis & Bregman, 1973; Lewis, Bregman, & Mahan, 1972) have demonstrated that established memories may be disrupted by ECS after a seven *day* retention interval provided that (retrieval) cues associated with the memory are reinstated just prior to the ECS administration. An analogous facilitation of reactivated memory was demonstrated by Gordon and Spear (1973).

### Memory is Not Simply the Performance of a Learned Act

Certain animal memory paradigms do not lend themselves to analyses of memory as performance of a learned act. Consider the following task (from Medin, 1969). On each trial monkeys see one cell of a 4 by 4 matrix of cells lights up, the light then goes out and, after a delay interval, the monkey must respond to the cell that was previously illuminated. The particular lighted cell changes randomly from trial to trial in this indirect delayed response procedure. No response is required (or allowed) during stimulus presentation.

It is hard to see how one would analyze this procedure in terms of performance of a learned act. The response (pushing open a cell door) made at the time of testing is clearly different from any response made during stimulus presentation. Not only that, but monkeys show immediate transfer to a task where two cells are lighted and two responses are allowed at the time of testing. Monkeys perform well on combinations of lights that they have never seen before (see Medin, 1969, Experiments 2-4). Analyzing performance in this task in terms of S-R associations could only be accomplished by means of an almost totally vacuous concept of stimulus, response, and association.

This observation is far from new. Tinklepaugh (1928, 1932) used a delayed response task in that chimpanzees would be shown where either lettuce or a banana was placed and then, after a delay, be allowed to choose among response sites. If the chimpanzee found the reward, it was allowed to consume it. Everything proceeded normally on trials when lettuce was hidden and was the reward, and when bananas were hidden and were the reward. Tinklepaugh also ran trials where bananas were hidden but then lettuce was surreptitiously substituted as the reward before the time of test. On these trials, the chimpanzees would select the correct location, but then with an expression that can only be described as puzzled (for quite convincing photographs see Tinklepaugh's 1928 article) they would ignore the lettuce and frantically search for the bananas that should have been there. Although bananas are preferred to lettuce and the design was not nicely counterbalanced, the implication is nonetheless clear. The chimpanzees were not just learning and remembering what to do, but also what the reward was.

### The Memory/Performance Distinction

Just as it has become common to distinguish between learning and performance, it is important to distinguish between memory and performance. To begin with, memory for an event may be expressed in a variety of ways and cannot simply be equated with a single index. Tinklepaugh's chimpanzees could manifest memory both in terms of making a correct response and in their reaction when the rewards were switched.

Equating memory with performance can lead to absurdities. For example, in a study evaluating memory in monkeys over very long delay intervals, D'Amato (personal communication, 1976) made a procedural change from using food pellets as the reward to using the more highly preferred reward of raisins. The immediate result was that performance fell essentially to chance! It seems that one would either have to concede a distinction between memory and performance or conclude that raisins are an amnesic agent. Actually, the latter possibility has to be entertained seriously in that Wagner and his associates have demonstrated that surprising events can disrupt stimulus processing (e.g., Terry & Wagner, 1975; Wagner, 1978, 1981; Wagner, Rudy & Whitlow, 1973). But if a switch in rewards is surprising, to describe this disruption a memory representation more complex than a stimulus-response association is needed and, to evaluate this more complex representation, more than one single underlying memory index is needed.

### Forgetting and Performance are Not a Simple Function of Associative Strength

One of the major ideas in contemporary memory research is the distinction between storage and retrieval. One major source of deficits in memory performance is retrieval failure (see Spear, 1973, 1976, 1978, 1981, for a very extensive review). The absence of correct responding does not mean that associative strength per se has been modified; simple changes in context and the corresponding absence of retrieval cues may account for poor performance. Excellent performance may return when the appropriate retrieval cues are reinstated.

An account of forgetting in terms of changes in associative strengths alone faces insurmountable problems. For example, there is now a considerable body of literature demonstrating that what are technically extinction trials can facilitate memory. Consider the following hypothetical experiment. Two groups of animals are trained in a conditioning task involving a simple CS-US associate. Following training we institute a two-month retention interval and then present the CS and record whether or not a CR appears. If one of the groups is put back in the test situation after one month and given a single extinction trial (CS but no US) that group can be expected to perform *better* than the other group on the later retention test. If the experiment group was presented with just the US and not the CS, we could expect a similar facilitation. Both procedures, from a learning theoretic perspective, should decrease associative strength and consequently lead to worse retention. They do not (see Campbell & Jaynes, 1966; Gordon, 1981; and Hamberg & Spear, 1978, for reviews of reactivation and reinstatement procedures).

## Summary

We have seen that there are serious problems with each of the assumptions associated with the learning approach to memory. Memory traces as well as stimulus traces may enter into associations. Memory is more than the performance of learned acts and it is apparent that one needs to distinguish between memory and performance. Finally, forgetting cannot be described adequately in terms of response competition or decay of associative strength.

The core of these difficulties is that this approach provides an impoverished view of memory. This framework is too simple, conceptually inadequate, and, not surprisingly, has fallen into disfavor. We turn now to what we take to be the current framework, which can be described as viewing learning from a memory perspective.

# THE PRESENT

## Attitudes

The observations that necessitate a broadened view of memory in animals reflect and encourage a more general perspective on memory research to the point that one can no longer provide a single set of assumptions to capture ongoing research on animal memory. For that reason, this section is more properly labeled as attitudes than assumptions.

Probably the most salient attitude is that constructs from ongoing research on human memory specifically, and cognitive psychology in general, deserve an airing in theories of animal memory. This does not mean that they are casually or automatically adopted; only that they are worth examining, much as human memory research has made ample use of constructs drawn from analogies to computers. It is probably far too early to evaluate this liberalized attitude in detail, but it can hardly be doubted that it has led to burgeoning research activity and corresponding thought.

A correlated attitude is that, in contrast to viewing memory from a learning perspective, one should turn the tables and view learning from a memory perspective. This serves to redress the imbalance noted earlier, although learning constructs may prove useful in analyzing such phenomena as improvements with practice on memory tasks.

A third attitude is that current memory research must be concerned not only with performance, but also the internal mechanisms and processes that govern performance. For example, rather than limiting itself to a unitary trace varying only in strength, contemporary research is concerned with the nature of memory representations, which can be quite varied, as well as the activities that operate on these representations. The following glimpse of contemporary research is admittedly sketchy, but it reveals much of the richness associated with this perspective.

## A Glimpse of Ongoing Research

### Influence of Cognitive Psychology on Animal Memory Research

*a. Short-Term Memory.*   The burst of research on human short-term memory in the 1960s has been followed by a similar wave of research on animal short-term memory in the 1970s. As interest in animal short-term memory continues, the number of stable empirical relationships uncovered steadily mounts. There is evidence from delayed-matching-to-sample (DMTS) studies, for example, that interference treatments alter memory for stimuli, responses, and reinforcers in a similar way (Maki, Moe, & Bierly, 1977) and produce the same pattern of results regardless of whether the subject population consists of pigeons, monkeys, or dolphins (D'Amato, 1974; Herman, 1975; Wilson, 1974; Zentall & Hogan, 1977). Associated with these empirical efforts are coordinated attempts to distinguish among alternative theories (D'Amato, 1974; Reynolds & Medin, 1981; Worsham, 1975).

Evidence that human short-term memory processes are flexible and depend on strategies such as rehearsal has stimulated corresponding interest in the flexibility of short-term memory in animals. Specifically, investigators are interested in the possibility that cueing animals to forget or remember some information will be reflected in memory performance (e.g., Grant 1981a, b; Maki, 1981; Rilling, Kendrick, & Stonebraker, 1982). The basic phenomenon can be demonstrated in animals but we feel that it is too early to conclude with certainty that animals have control processes on the order of flexibility attributed to human subjects.

*b. List Memory.* Studies of human memory almost always use more than one to-be-remembered item or event and corresponding animal memory tasks have yielded intriguing parallels (e.g., Davis & Fitts, 1976; Eddy, 1973; Gaffan, 1977; Roberts & Kraemer, 1981; Sands & Wright, 1980a, b; Thompson & Herman, 1977). For example, using a probe recognition technique, Sands and Wright have demonstrated both recency *and* primacy effects in list memory. These findings are of particular interest in that some memory researchers have assumed that recency effects derive from a (verbally-based) rehearsal buffer and that primacy effects can be attributed to differential verbal rehearsal.

*c. Retrieval.* The distinction between storage and retrieval processes and the concept of retrieval cues, so prominent in human memory research, has been developed and elaborated to account for a very wide range of forgetting phenomena in animal memory (e.g., Miller & Springer, 1973; Spear 1978). There is abundant evidence showing that changes in context may impair memory and that procedures for reactivating or reinstating contextual cues can produce a significant facilitation of retention in animals.

Retrieval processes even influence processing in short-term memory tasks. For example, in a DMTS paradigm we have found that an interpolated stimulus similar to the to-be-remembered stimulus along only irrelevant dimensions can produce facilitation in memory performance (Medin, Reynolds, & Parkinson, 1980). One interpretation of these results is that the interpolated stimulus acts as a retrieval cue leading to the activation and further processing of the initial stimulus.

### Learning From a Memory Perspective

It has proven quite fruitful to draw on developments in memory theories to aid in the formulation of more adequate models of conditioning and learning. To list a small number of examples, this has been evidenced in the areas of the learning of sequences (e.g., Capaldi, 1971), Pavlovian conditioning (e.g., Estes, 1973; Wagner, 1981; Wagner, Rudy, & Whitlow, 1973), taste aversion learning (Milgram, Krames, & Alloway, 1977; Revusky, 1971), learning with delayed reinforcement (D'Amato & Cox, 1976), discrimination learning (Medin, 1975), and serial reversal learning (Williams, 1976).

The well-known Rescorla-Wagner theory of Pavlovian conditioning (Rescorla & Wagner, 1972) carries with it the implication that a CS-UCS episode will promote learning only to the degree that the UCS is surprising or unexpected. Wagner and his associates (e.g., Wagner, Rudy, & Whitlow, 1973) have examined and gathered strong support for the idea that a surprising UCS initiates a posttrial rehearsal process that is necessary for associative learning. An important factor in this work is that rehearsal is clearly defined in a manner that suggests further lines of inquiry into processing of episodes (such as, for example, analyses of backward conditioning in terms of the temporal patterns of CS and UCS rehearsal). This list could go on.

Recently several investigators have suggested that the phenomena of blocking and overshadowing where learning about one cue interferes with learning about another cue is, in part, a retrieval failure problem and that learning about the second cue can be demonstrated when that retrieval problem is circumvented (Balaz, Gutsin, Cacheiro, & Miller, 1982; Kasprow, Cacheiro, Balaz, & Miller, 1982; Kaufman & Bolles, 1981).

### Representation and Processing

*a. Attributes of Memory.* In both human and animal memory research, the idea that traces are unitary has been replaced by the idea that a memory trace may be comprised of a collection of attributes (e.g., Bower, 1967; Spear, 1978; Underwood, 1969; Wickens, 1970). The implications of the idea of alternative memory attributes are just starting to be exploited. Specific memory models are frequently based on the assumption that memory performance derives from a single underlying attribute.

One may also note that there is no guarantee that different attributes will follow the same principles. Consider the literature on a basic phenomenon, the effects of spacing of repetitions on memory. To say the least, the animal literature presents a rather confusing picture. Spaced repetitions sometimes facilitate memory (e.g., Medin, 1974; Robbins & Bush, 1973; Roberts, 1974) and sometimes impair it (e.g., Herzog, Grant, & Roberts, 1977; Medin, 1974, 1980; Roberts, 1972). One possibility is that spacing facilitates memory for some attributes but not for others. For example, a suggestive pattern in the repetition spacing literature is that spaced repetitions seem to help only in those circumstances where familiarity may play a role in performance.

The attributes controlling performance may also shift as a function of an animal's experience at a task. For example, a recent series of studies in our laboratory looking at the role of stimulus familiarity in DMTS performance by monkeys reveals one such shift (Parkinson & Medin, 1982). Four trial types were created by having the sample stimulus be either relatively novel and rewarded $(N^+)$ or familiar and rewarded $(F^+)$, in conjunction with a novel $(N^-)$ or familiar $(F^-)$ unrewarded foil at time of test. Early in training, the trial types yielded the following order: performance on $F^+N^-$ trials was better than $N^+N^-$, which was better than $F^+F^-$, which surpassed $N^+F^-$. This pattern of results is consistent with the idea that monkeys prefer familiar stimuli but that familiarity does not otherwise affect performance. As the animals received more training, however, the pattern of performance shifted, and the ordering became $F^+N^- > N^+F^- > N^+N^- > F^+F^-$. We believe that this pattern of performance can best be understood by the idea that familiarity and novelty came to be used as discriminative attributes. Quite apart from any influences on preference, the monkeys seemed to be able to use the fact that a given sample was novel (familiar) to reject incorrect alternatives that were familiar (novel).

*b. Coding Processes.*    Investigators have also been concerned with the form in which information is stored by animals in memory tasks (e.g., Honig, 1978, 1981; Honig & Thompson, 1982; Roitblat, 1980, 1982). For example, Roitblat (1980) used a symbolic matching to sample task (e.g., a red sample means choose the triangle at the time of test) to ask whether pigeons encode information retrospectively (that is, do they try to remember the sample) or prospectively (do they try to remember which stimulus they should choose at the time of test). It would be naive to expect a simple answer to this question and it is probably more appropriate to ask when one type of encoding will predominate, but, for the record, Roitblat's three birds encoded prospectively.

Another recent line of research concerns an animal's ability to generate complex temporal sequences of behavior (e.g., Straub, 1979; Straub, Seidenberg, Bever, & Terrace, 1979; Straub & Terrace, 1981; Terrace, Straub, Bever, & Seidenberg, 1977). Pigeons can learn to peck at least four colored keys in correct sequence (e.g., blue, green, red, yellow) regardless of their spatial position. Paradigms of this type are leading to ever more sophisticated models of pigeon memory (e.g., Straub, et al., 1979) and sharper questions concerning which types of representation are compatible with the resulting patterns of performance.

## Summary

The previous topics reflect only a portion of current research in animal memory and even less of the excitement and theoretical interest associated with it. It is hard to escape the impression that changing attitudes have led to substantial progress. Therefore, it is with a note of apology that we point toward obstacles that we see ahead rather than applaud current gains. In the next paragraphs three major issues are outlined that we think will have to be addressed if progress in understanding animal memory is to continue.

## Problems and Limitations

### Units of Analysis

If a large, red triangle is presented as a sample in a delayed matching to sample trial, are we asking the animal to remember one thing (the stimulus) or three (size, color, shape)? Studies of human memory suggest that the functional unit can shift depending on the components. For example, in a short-term memory task, people can remember three words as readily as they can remember three unrelated consonants (Murdock, 1961).

What are the proper (functional) units of analysis in animal memory experiments? How can one specify memory load without knowing these units? Does the compounding of a color and a form in a memory experiment act fundamentally the same or different from the compounding of a color and a tone? Unless we know more about organizational units in animal memory, empirical relationships may be misconstrued and theories may be found wanting, not because they specify the processes incorrectly but because the units that the processes operate on have been misspecified. (See Shimp, 1976, for additional arguments concerning units of analysis.) In brief, units of analysis looms as a major, unanswered question.

### Categorical Memory

One consequence of taking the individual memory episode as the fundamental unit of analysis is that very little is known about memories which logically must be derived from the integration of a number of specific episodes. In studies of human memory it has proven useful to draw a distinction between episodic and semantic memory (Tulving, 1972). *Episodic memory* refers to memory for specific, dated events or episodes such as one's most recent phone conversation. *Semantic memory* refers to symbolic information, abstract rules or procedures, or knowledge of meaning that is not tied to any one particular episode. Examples might be one's knowledge of word meanings or knowing in general how phones are used. Because the term *semantic memory* has the specific connotation of word meaning we will use the term *categorical memory* to refer to general knowledge of rules, procedures or significances that are based on the integration of multiple episodes or experiences. It is surprising that the recent explosion of interest in semantic memory, which underlies much of cognitive psychology, has not yet been accompanied by a corresponding interest in the integration of information in animal memory.

The neglect of categorical memory represents both a limitation and a missed opportunity. The memory procedures used in the laboratory often are based on months of preliminary training during which the animals develop the expertise needed to perform adequately in the experiments. To the extent that our theories focus on memory for the specific information that is the substance in a particular study and ignore the contributions of categorical memory to performance in the task, our theories are severely limited. This focus on single episode is also a limitation because there is increasing evidence that memories may interact with one another. Numerous studies indicate that a newly presented stimulus can either interfere with or facilitate memory for an old stimulus depending on factors such as the new stimuli either retrieving memory for the older one or displacing the older one from short-term memory (e.g., see Spear & Miller, 1981).

Investigators working on human memory face several constraints in studying semantic memory. They usually have subjects available for only an hour or so and consequently typically study performance based on semantic memory rather than the acquisition, integration, and incorporation of new knowledge into semantic memory. Researchers studying animal memory have a unique advantage in that the subjects can be trained for extended periods of time in well-controlled circumstances. In principle, we have the opportunity to study the development of categorical memory in a manner that is rarely practical in human studies. Unfortunately, investigators rarely have exploited this opportunity to study the development of categorical memory in animals.

### Cognitive Ethology of Memory

Relatively little attention has been directed at the particular stimuli used in animal memory experiments from the point of view of the evolutionary history or special adaptations of the species in question. We think this circumstance is directly analogous to studying human memory with the constraint that the stimulus materials be nonsense syllables. That constraint was a common practice at an early stage of analyses of human memory, but it is hardly satisfactory in current times.

If studies of animal memory showed more of a bent toward cognitive ethology, then we might know more about proper units of analysis, stimulus meaningfulness, and could use more natural categories in studies of categorical memory. Some psychologists have argued that human memory can best be understood as a byproduct of other cognitive activities. If this is so, it may be a mistake to study animal memory divorced from other cognitive activities. Ethologically motivated analyses may tell us more about some of these activities.

## PROSPECTS AND PROJECTIONS

In this section our aim is to back up our points by discussing the three issues outlined previously in detail and illustrating them, where possible, with examples from ongoing research.

### Units of Analysis

It is becoming increasingly obvious that theorizing about animal learning and memory will be heavily influenced by the particular units of analysis selected. Typically, experimenters analyze complex stimuli into components, but little is known about whether and how animals do.

### Memory Load

There is abundant evidence that human short-term memory varies as a function of load—it is far easier to hold one item of information in STM than four (e.g., Crowder, 1976). Interestingly, there is evidence that pigeons remember single stimuli (e.g., red, circle) better than a compound (red circle) comprised of two stimuli (Maki & Leith, 1973; Maki, Riley, & Leith, 1976). This result was interpreted in terms of memory load, but Roberts and Grant (1978) and Cox and D'Amato (1982) ran analogous experiments and argued that the best interpretation of the data was that the compound stimuli were not analyzed into components and that differences in performance on elements and compounds arose from stimulus generalization decrements. It would be premature to try to resolve these different interpretations. However, one cannot analyze memory load unless one knows what the elementary units of the to-be-remembered material are. This is especially relevant to theories of animal memory incorporating the notion of memory span as a construct (see the chapter by Kendrick & Rilling, this volume).

### Integral Versus Separable Dimensions

Work on human information processing has led to the distinction between integral and separable dimensions (e.g., Garner, 1974). A number of converging operations underly the distinction, but basically the difference lies in whether or not selective attention to one dimension of a pair is possible. For example, college students find size and shape to be separable dimensions and can rapidly sort stimuli by size while ignoring shape. Hue and brightness act as integral dimensions and selective attention is much more difficult.

Lamb and Riley (1981) examined integral and separable dimension in pigeon delayed matching-to-sample. Integral and separable dimensions were defined in terms of spatial proximity. The elements used were lines and colors and integral stimuli were colored lines whereas separable stimuli had color and line aspects in spatially distinct locations. The different arrangements produced marked differences in performance. Integral stimuli were matched as well as elements, but separable stimuli were associated with decrements in performance. Although these effects of stimulus arrangement were large, their interpretation is unclear in the absence of converging evidence on whether the dimensions were, in fact, integral and separable for the birds. Dimensional processing currently is of great interest in developmental psychology (e.g., Kemler, 1982; Shepp, 1978; Smith & Kemler, 1978), and this literature makes it clear that integrality and separability are not fixed stimulus properties but change with experimental variables and experience. For example, there is a general development from dimensions acting as integral to their acting as separable. It would be interesting to see if animals also would show a shift in dimensional processing as a function of experience.

### Independence of Stimulus Components

Early models of discrimination learning (e.g., Spence, 1936) assumed that associative strength became independently attached to the components of compound stimuli. This assumption credits animals with the ability to analyze stimuli into components in a way that implies excellent powers of abstraction.

An alternative discrimination learning model proposed by one of us (Medin, 1975) rejects the assumption of independence of components. The main idea is that each component provides a context for other components and that for information about a component to be accessed, both the cue and its associated context must be activated simultaneously. For example, what an animal learns about *red* when it appears on a triangle may not transfer to *red* when it appears on a circle, if the change in context associated with switching from triangle to circle is too great. That is, components combine in an interactive rather than an independent manner. Some direct attempts to contrast independent and interactive models in discrimination and classification learning have yielded data favoring interactive cue models (Medin, 1975; Medin & Schaffer, 1978; Medin & Smith, 1981).

More recently this context model has been applied to analyses of proactive interference in delayed-matching-to-sample performance by monkeys (Reynolds & Medin, 1981). Between-trial proactive interference was studied in a situation in which the similarity of consecutive trials was varied along the dimensions of color, form, and position. All of these factors, as well as the similarity of sample and test contexts, contributed to memory performance. A mathematical model assuming an interaction of component dimensions gave an excellent qualitative and quantitative account of the data in each of two experiments. Certain counterintuitive predictions, such as that moving the choice stimuli closer together would facilitate performance, were supported. Independent cue models failed to describe these data.

Although the previous observations are consistent with the general idea that components are not encoded independent of one another, many questions remain concerning the nature of this nonindependence. For example, one might imagine that a compound of red and triangle may be linked more tightly than a compound of a light and a tone. As another example, Rescorla and Durlach (1981) have given up the notion of component independence based on their evidence that associations develop between elements of a compound (e.g., flavor plus smell). Further, these associations can dramatically modify typical blocking and overshadowing effects.

These examples are designed to reinforce the argument that there is a need to know more about appropriate units in analyzing animal memory. We have hinted that the units may change with experience and that is just one of several interesting problems that should be explored. For example, little is known about how components do or do not interact in forgetting. In addition, there is evidence that in some situations performance with respect to a compound stimulus is better than would be expected on the basis of performance on components. Medin (1969) presented monkeys with patterns consisting of the illumination of one to four cells of $4 \times 4$ matrix of cells and found that differences in performance on some four-cell patterns (e.g., those forming a square) versus others (four random cells) could not be explained in terms of performance on the individual cells when from single light trials.

By now the point should be clear. Not only does the absence of knowledge of units of analysis represent an obstacle to memory research, but there is also more than a hint that attention to the issue of units will uncover new problems and relationships (e.g., within compound associations) that will advance our understanding of animal learning and memory.

## Categorical Memory

Although we are arguing that investigators have tended to ignore categorical memory in favor of studying memory for a single event, there are enough examples available from current research to illustrate some intriguing issues associated with categorical memory.

### Conceptual Relationships

Studies of concept learning in animals can be thought of as involving or requiring categorical memory because appropriate performance requires integrating information from a series of events or episodes. Actually, there is a fairly extensive research literature on concept learning in animals (see French, 1965, and Stone, 1951 for sample reviews) conducted mainly from the point of view of establishing the limitations and capacities of the organisms in question.

Earlier work on capacities is being supplemented with some interesting and more detailed analyses of cognitive performance. For example, Herrnstein and his associates (Cerella, 1979; Herrnstein, 1979; Herrnstein & DeVilliers, 1981; Herrnstein, Loveland, & Cable, 1976) have trained pigeons to respond to slides having in common only the property of depicting concepts like water, trees, fish, or a particular person. What is especially interesting is that even though there seems to be no single set of properties that slides of the concept must have, pigeons nonetheless seem to learn these ill-defined concepts as rapidly as they learn more standard laboratory tasks that do have common features.

One of the main payoffs of the controversial studies of language learning in great apes (e.g., Gardner & Gardner, 1978; Premack, 1976; Rumbaugh, 1977; Terrace, 1979) is that we may learn much more about animal conceptual behavior. Many of the tests for "language" involve the teaching and performance of abstract concepts such as sameness versus difference (see also Zentall, Edwards, Moore, & Hogan, 1981).

### Episodic-Categorical Relationship

In addition to the often mysterious process by which a set of episodes or examples leads to the development of general knowledge or categorical memory, there are important unanswered questions about the relationships between episodic and categorical memory. Consider, for example, discrimination learning set or learning to learn by monkeys. When monkeys are given a small number of trials on many different problems, they start out learning very slowly but eventually they come to master new discriminations in a single trial. This remarkable ability does not develop if a large number of trials is given on a single discrimination, and repeated reversals of a single discrimination are less effective than the same number of trials distributed over a variety of discrimination problems (see Medin, 1977, for a review). One set of questions concerns just what episodic events are necessary and sufficient for learning set to develop.

A second set of interesting questions concerns how *memory* for a single discrimination changes as learning set develops. One might expect that as more problems are learned there should be more proactive interference and memory for a particular discrimination should suffer. Although the proper way to measure retention is not a neutral issue, there is evidence that memory for particular problems does decrease as the learning set develops, but that with further training memory improves substantially (Medin, 1977). This suggests that in parallel with learning to learn there may be the development of learning to remember.

There are studies of long term memory for general skills and abstract concepts (e.g., Patterson & Tzeng, 1979; Stollnitz, 1970). These skills prove to be resistant to forgetting even when retention intervals are on the order of years.

### Cognitive Maps

Most readers may be acquainted with the observation that animals may show very impressive spatial memory. For example, rats can simultaneously remember which of 18 alleys they have and have not entered (e.g., Olton, 1978). That is, they can retrieve a reward from each alley without duplicating their responses and show little interference across several sets of 18 runs.

Menzel (1978) has noted similar spatial skills in chimpanzees. He has shown that chimpanzees exposed to rewards at various locations in a haphazard manner will retrieve those rewards not in the order in which they were presented but rather in a systematic manner based on their spatial locations. Observations such as this suggest that chimpanzees are capable in integrating a series of events into a unified cognitive map.

The previous examples are not exhaustive. We think that the work involving interactions among memories discussed in this chapter will naturally lead to research on categorical memory and episodic categorical relationships.

## Cognitive Ethology of Memory

Just as human memory for meaningful material is vastly better than memory for nonsense syllables, our appreciation of memory ability (and processes) in animals may be strongly limited by the nature of the stimulus materials that typically are employed. The impressive spatial memory that some species of animals display must derive in part from the fact that experimenters have hit upon a task that is "natural" for them. This is just one of the virtues that may be associated with using stimulus materials that have more meaning and importance to the organism.

### Spatial Memory

Excellent spatial memory is not confined to mammals. Jumping gobies are a species of fish inhabiting tidal pools, and when they return to the sea they have to jump from pool to pool. Because they cannot see from one pool to another, the gobies must remember where the various pools were from swimming around during high tide. Because tidal pools are different in different locations and shift over time, the gobies' behavior provides a natural memory experiment that could be brought into the laboratory. In the same vein, Balda and Turek (1982) have exploited the seed burying behavior of nutcrackers to study memory in these birds. These studies are of interest to investigators of animal memory, because well-developed skills in natural tasks may reveal memory processes (e.g., rehearsal) that would only be weakly evidenced in less meaningful contexts.

### Animal Communication

Marler and Peters (1981) showed that a bird can be exposed to the song for its species before it is able to sing and retain the pattern over a period of months. That memory acts as a template that the bird uses to develop and fine tune its song when it later begins to sing. The range of adaptations in different song birds has led to a range of flexibility in learning, a range that could be exploited in memory experiments. In addition, recent work on species specific processing of vocal sound by monkeys (Zoloth, Petersen, Beecher, Green, Marler, Moody, & Stebbins, 1979) suggests interesting possibilities for research on categorical memory in primates. To give but one example, an important issue in human categorization research concerns whether memory for a category is based on memory for particular instances or on a more abstract summary representation (e.g., a mental average or prototype). It may be that monkeys use one form of representation for vocalizations of their own species and a different one for vocalizations from conspecific monkeys. Alternatively, one form or representation may only hold for vocalizations that have significance for a monkey (e.g., alarm calls versus more random vocalizations.)

### Face Perception

Although one might not think of pictures as being very natural stimuli, there is evidence that colored slides of monkeys can trigger innate releasing mechanisms in infant monkeys (Sackett, 1966). For example, pictures of a monkey making a "threat" gesture disturbed the behavior of infant monkeys that had been reared in isolation from birth. Monkeys can discriminate individual faces of other monkeys and are not confused by manipulations in orientation, posture, size, color, or illumination (Rosenfeld & Van Hoesen, 1979). Using techniques borrowed from studies of infant perception, Humphreys (1972, 1974) has studied changes in how pictures are treated by monkeys. For example, inexperienced animals treat individual monkeys as being quite different and individual domestic animals of the same species as being closely similar. Experienced monkeys, exposed to pictures over a period of six months, treat all individuals as different from each other. These observations suggest that pictures of faces would be excellent stimuli for studies of both episodic and categorical memory.

The motivation for using more natural stimuli and tasks is not to produce better memory as an end in itself. There is a long history in comparative research of trying to study a phenomenon in its most developed and accessible form. The size of squid axons commends them to study, for example. Similarly, we may gain better insight into memory processes in general by using stimuli and tasks that bring out the best in organisms.

## Summary

The reader can readily perceive that the three themes discussed as problems and prospects actually fit together quite closely. By using natural stimuli, we are more likely to hit upon the appropriate units of analysis. Understanding units of analysis will allow greater insight in evaluating relationships between episodic and categorical memory. Focusing on aspects of categorical memory that are natural to organisms promises to be a shortcut to developing better and more general theories of animal memory.

The reader may also have noticed that there is an implicit or explicit competition among the traditional learning orientation, cognitive psychology, and ethology. When considered from the perspective of meta-assumptions and preferences in theorizing, these perspectives may appear to be mutually exclusive. Yet, we think that each viewpoint has much to offer the others. Cognitive psychology has profited greatly by rejecting the limited view of memory associated with the traditional learning approach to memory. At the same time, however, the growing interest in the nature of expertise and the acquisition of skills is leading to a renewed recognition of the importance of learning on the part of cognitive psychologists. The backdrop of several decades of learning research will doubtless facilitate attempts to incorporate learning processes into cognitive processes. One also sees within cognitive psychology a narrowing of the gap between theoretical and applied research that is proving to be quite fruitful. The standard concept learning tasks are being supplemented by studies of thought processes and decision in complex tasks (e.g., medical problem solving, managing a nuclear power plant). These studies raise new questions (e.g., how does decision making change when the information is presented sequentially versus all at once) and sharpening theoretical questions. We think that an ethological perspective offers animal memory researchers the same challenge and stimulation that applied problems have offered cognitive psychologists working with people. To address the issues we have outlined as crucial for further advances in our understanding of animal memory, investigators will need to draw on traditional learning approaches, cognitive psychology, and ethology.

## A FINAL NOTE

It should be clear that we have been painting a picture of progress. Viewed against the backdrop of research of only a few decades ago, substantial advances have been both empirically and theoretically. But continued progress does not automatically follow on the heels of current successes, and we think that future analyses of animal memory will be gauged, in part at least, by how well they address the issues that have been outlined in this paper.

## ACKNOWLEDGMENTS

This research was supported by National Science Foundation Grant BNS 79-22678. Linda Powers provided helpful comments on an earlier draft of the manuscript.

## REFERENCES

Anderson, J. R. *Language, memory and thought*. Hillsdale, NJ: Lawrence Erlbaum Associates, 1976.
Anderson, J. R. Interference: The relationship between response latency and response accuracy. *Journal of Experimental Psychology: Human Learning and Memory*, 1981, 7, 326–343.
Anderson, J. R., & Bower, G. *Human associative memory*. New York: Winston, 1973.
Balaz, M. A., Austin, P., Cacheiro, H., & Miller, R. R. Blocking as a retrieval failure: Reactivation of associations to a blocked stimulus. *Quarterly Journal of Experimental Psychology*, 1982, in press.
Balda, R. P., & Turek, R. J. The cache-recovery system as an example of memory capabilities in Clark's nutcracker. Paper presented at the Guggenheim Conference on Animal Cognition, Columbia University, 1982.
Bolles, R. C. Learning, motivation and cognition. In W. K. Estes (Ed.), *Handbook of learning and cognitive processes*. Vol. L. Hillsdale, NJ: Lawrence Erlbaum Associates, 1975.
Bolles, R. C. Some relationships between learning and memory. In D. L. Medin, W. A. Roberts, & R. T. Davis (Eds.), *Processes of animal memory*. Hillsdale, NJ: Lawrence Erlbaum Associates, 1976.

Bower, G. M. A multicomponent theory of the memory trace. In K. W. Spence & J. T. Spence (Eds.), *The psychology of learning and motivation: Advances in research and theory*, Vol. 1. New York: Academic Press, 1967.

Campbell, B. A., & Jaynes, J. Reinstatement. *Psychological Review*, 1966, *73*, 478–480.

Capaldi, E. J. Memory and learning: A sequential viewpoint. In W. K. Honig & P. H. R. James (Eds.), *Animal memory*. New York: Academic Press, 1971.

Cerella, J. Visual classes and natural categories in the pigeon. *Journal of Experimental Psychology: Human Perception and Performance*, 1979, *5*, 68–77.

Cox, J. K., & D'Amato, M. R. Matching to compound samples by monkeys *(Cebus apella)*: Shared attention or generalization decrement? *Journal of Experimental Psychology: Animal Behavior Processes*, 1982, *8*, 209–225.

Crowder, R. G. *Principles of learning and memory*. Hillsdale, NJ: Lawrence Erlbaum Associates, 1976.

D'Amato, M. R. Delayed matching and short-term memory in monkeys. In G. H. Bower (Ed.), *The psychology of learning and motivation: Advances in research and theory*, Vol. 7. New York: Academic Press, 1973.

D'Amato, M. R., & Cox, J. K. Delay of consequences and short term memory in monkeys. In D. L. Medin, W. A. Roberts, & R. T. Davis, *Processes in animal memory*. Hillsdale, NJ: Lawrence Erlbaum Associates, 1976.

D'Amato, M. R., Safarjan, W. R., & Salmon, D. Long-delay conditioning and instrumental learning: Some new findings. In N. E. Spear & R. R. Miller, (Eds.), *Information processing in animals: Memory mechanisms*. Hillsdale, NJ: Lawrence Erlbaum Associates, 1981.

Davis, R. T., & Fitts, S. S. Memory and coding processes in discrimination learning. In D. L. Medin, W. A. Roberts, & R. T. Davis (Eds.), *Processes in animal memory*. Hillsdale, NJ: Lawrence Erlbaum Associates, 1976.

Eddy, D. R. Memory processing in *Macaca speciosa:* Mental processes revealed by reaction time experiments. Unpublished doctoral dissertation, Carnegie-Mellon University, 1973.

Estes, W. K. Memory and conditioning. In F. J. McGuigan & D. B. Lunsden (Eds.), *Contemporary approaches to conditioning and learning*. New York: Wiley, 1973.

French, G. M. Associative problems. In A. M. Schrier, H. F. Harlan, & R. Stollnitz (Eds.), *Behavior of nonhuman primates: Modern research trends*, Vol. 1. New York: Academic Press, 1965.

Gaffan, D. Recognition memory after short retention intervals in fornix-transected monkeys. *Quarterly Journal of Experimental Psychology*, 1977, *29*, 577–588.

Garcia, J., Ervin, F. R., & Koelling. Learning with prolonged delay of reinforcement. *Psychonomic Science*, 1966, *5*, 121–122.

Gardner, R. A., & Gardner, B. T. Comparative psychology and language acquisition. *Annals of the New York Academy of Sciences*, 1978, *309*, 37–76.

Garner, W. R. *The processing of information and structure*. Potomac, MD: Lawrence Erlbaum Associates, 1974.

Gordon, W. C. Mechanisms of cue-induced retention enhancement. In N. E. Spear & R. R. Miller (Eds.), *Information processing in animals: Memory mechanisms*. Hillsdale, NJ: Lawrence Erlbaum Associates, 1981.

Gordon, W. C., & Spear, N. E. The effects of strychnine on recently acquired and reactivated passive avoidance memories. *Physiology and Behavior*, 1973, *10*, 1071–1075.

Grant, D. S. Short-term memory in the pigeon. In N. E. Spear & R. R. Miller (Eds.), *Information processing in animals: Memory mechanisms*. Hillsdale, NJ: Lawrence Erlbaum Associates, 1981(a).

Grant, D. S. Intertrial interference in rat short-term memory. *Journal of Experimental Psychology: Animal Behavior Processes*, 1981(b), *7*, 217–227.

Hamberg, J. M., & Spear, N. E. Alleviation of forgetting of discrimination learning. *Learning and Motivation*, 1978, *9*, 466–476.

Herman, L. M. Interference and auditory short-term memory in the bottlenosed dolphin. *Animal Learning and Behavior, 1975, 3,* 43–48.

Herrnstein, R. J. Acquisition, generalization, and discrimination reversal of a natural concept. *Journal of Experimental Psychology: Animal Behavior Processes, 1979, 5,* 116–129.

Herrnstein, R. J., & DeVilliers, P. A. Fish as a natural category for people and pigeons. In G. H. Bower (Ed.), *The psychology of learning and motivation,* Vol. 14. New York: Academic Press, 1980.

Herrnstein, R. J., Loveland, D. H., & Cable, C. Natural concepts in pigeons. *Journal of Experimental Psychology: Animal Behavior Processes, 1976, 2,* 285–311.

Herzog, H. L., Grant, D. S., & Roberts, W. A. Effects of sample duration and spaced repetition upon delayed matching to sample in monkeys (Macaca arctoides and saimiri scirueus). *Animal Learning and Behavior, 1977, 5(4),* 347–354.

Honig, W. K. Studies of working memory in the pigeon. In S. H. Hulse, H. Fowler, and W. K. Honig (Eds.), *Cognitive processes in animal behavior.* Hillsdale, NJ: Lawrence Erlbaum Associates, 1978.

Honig, W. K. Working memory and the temporal map. In N. E. Spear & R. R. Miller (Eds.), *Information processing in animals: Memory mechanisms.* Hillsdale, NJ: Lawrence Erlbaum Associates, 1981.

Honig, W. K., & Thompson, R. K. R. Retrospective and prospective processing in animal working memory. In G. H. Bower (Ed.), *The psychology of learning and motivation,* Vol. 16. New York: Academic Press, 1982, in press.

Humphrey, N. K. 'Interest' and 'pleasure': Two determinants of a monkey's visual preference. *Perception, 1972, 1,* 395–416.

Humphrey, N. K. Species and individuals in the perceptual world of monkeys. *Perception, 1974, 3,* 105–114.

Kasprow, W. J., Cacheiro, H., Balaz, M. A., & Miller, R. R. Reminder-induced recovery of associations to an overshadowed stimulus. *Learning and Motivation, 1982,* in press.

Kaufman, M. A., & Bolles, R. C. A nonassociative aspect of overshadowing. *Bulletin of the Psychonomic Society, 1981, 18,* 318–320.

Kemler, D. G. Wholistic and analytic modes in perceptual and cognitive development. In T. Tighe & B. Shepp (Eds.), *Interactions; Perception, cognition and development.* Hillsdale, NJ: Lawrence Erlbaum Associates, 1982.

Lamb, M. R., & Riley, D. A. Effects of element arrangements on the processing of compound stimuli in pigeons (Columba livia), *Journal of Experimental Psychology: Animal Behavior Processes, 1981, 7,* 45–58.

Lewis, D. A cognitive approach to experimental amnesia. *American Journal of Psychology, 1976, 89,* 51–80.

Lewis, D. J., & Bregman, J. H. The source of the cues for cue dependent amnesia. *Journal of Comparative and Physiological Psychology, 1973, 85,* 421–426.

Lewis, D. J., Bregman, J. J., & Mahan, J. J. Cue-dependent amnesia in rats. *Journal of Comparative and Physiological Psychology, 1972, 81,* 243–247.

Maki, W. S. Directed forgetting in animals. In N. E. Spear & R. R. Miller (Eds.), *Information processing in animals: Memory mechanisms.* Hillsdale, NJ: Lawrence Erlbaum Associates, 1981.

Maki, W. S., & Leith, C. R. Shared attention in pigeons. *Journal of the Experimental Analysis of Behavior, 1973, 19,* 345–349.

Maki, W. S., Moe, J. C. & Bierly, C. M. Short-term memory for stimuli, responses, and reinforcers. *Journal of Experimental Psychology: Animal Behavior Processes, 1977, 3,* 156–177.

Maki, W. S., Riley, D. A., & Leith, C. R. The role of test stimuli in matching to compound samples by pigeons. *Animal Learning and Behavior, 1976, 4,* 13–21.

Marler, P., & Peters, S. Sparrows learn adult song and more from memory. *Science*, 1981, *213*, 780–782.

Medin, D. L. Form perception and pattern reproduction in monkeys. *Journal of Comparative and Physiological Psychology*, 1969, *68*, 412–419.

Medin, D. L. The comparative study of memory. *Journal of Human Evolution*, 1974, *3*, 455–463.

Medin, D. L. A theory of context in discrimination learning. In G. H. Bower (Ed.), *The psychology of learning and motivation*, Vol. 9. New York: Academic Press, 1975.

Medin, D. L. Memory processes and discrimination learning set formation. In A. M. Schrier (Ed.), *Progress in Behavioral Primatology*. Hillsdale, NJ: Lawrence Erlbaum Associates, 1977.

Medin, D. L. Proactive interference in monkeys: Delay and intersample interval effects are noncomparative. *Animal Learning and Behavior*, 1980, *8(9)*, 553–560.

Medin, D. L., Reynolds, T. J., & Parkinson, J. K. Stimulus similarity and retroactive interference and facilitation in monkey short term memory. *Journal of Experimental Psychology: Animal Behavior Processes*, 1980, *6(2)*, 112–125.

Medin, D. L., & Schaffer, M. M. Context theory of classification learning. *Psychological Review*, 1978, *85*, 207–238.

Medin, D. L., & Smith, E. E. Strategies and classification learning. *Journal of Experimental Psychology: Human Learning and Memory*, 1981, *7*, 241–253.

Menzel, E. W. Cognitive mapping in chimpanzees. In S. M. Hulse, A. Fowler, & W. K. Honig (Eds.). *Cognitive processes in animal behavior*. Hillsdale, NJ: Lawrence Erlbaum Associates, 1978.

Miller, R. R., & Springer, A. D. Amnesia, consolidation and retrieval. *Psychological Review*, 1973, *80*, 69–79.

Murdock, B. B., Jr. The retention of individual items. *Journal of Experimental Psychology*, 1961, *62*, 618–625.

Olton, D. S. Characteristics of spatial memory. In S. H. Hulse, H. Fowler, & W. K. Honig (Eds.), *Cognitive processes in animal behavior*. Hillsdale, NJ: Lawrence Erlbaum Associates, 1978.

Parkinson, J. K., & Medin, D. L. Emerging attributes in monkey short-term memory. *Journal of Experimental Psychology: Animal Behavior Processes*, 1982, in press.

Patterson, T. L., & Tzeng, O. J. L. Long-term memory for abstract concepts in the lowland gorilla *(Gorilla g. Gorilla)*. *Bulletin of the Psychonomic Society*, 1979, *13*, 279–282.

Premack, D. A. *Intelligence in ape and man*. Hillsdale, NJ: Lawrence Erlbaum Associates, 1976.

Rescorla, R. A., & Durlach, P. J. Within-event learning in Pavlovian conditioning. In N. E. Spear & R. R. Miller (Eds.), *Information processing in animals: Memory mechanisms*. Hillsdale, NJ: Lawrence Erlbaum Associates, 1981.

Rescorla, R. A., & Wagner, A. R. A theory of Pavlovian conditioning: Variations in the effectiveness of reinforcement and nonreinforcement. In A. H. Black & W. F. Prokasy (Eds.), *Classical conditioning II*. New York: Appleton-Century-Crofts, 1972.

Revusky, S. The role of interference in association over a delay. In W. K. Honig & P. H. R. James (Eds.), *Animal memory*. New York: Academic Press, 1971.

Reynolds, T. J., & Medin, D. L. Stimulus interaction and between-trial proactive interference in monkeys. *Journal of Experimental Psychology: Animal Behavior Processes*, 1981, *7*, 334–347.

Rilling, M., Kendrick, D. F., & Stonebraker, T. B. Stimulus control of forgetting: A behavioral analysis. In M. L. Commons, A. R. Wagner, & R. J. Herrnstein (Eds.), *Quantitative studies in operant behavior: Acquisition*. Cambridge, MA: Ballinger, 1982.

Robbins, D., & Bush, C. T. Memory in great apes. *Journal of Experimental Psychology*, 1973, *97*, 344–348.

Roberts, W. A. Short-term memory in the pigeon: Effects of repetition and spacing. *Journal of Experimental Psychology,* 1972, *94,* 74–83.

Roberts, W. A. Spaced repetition facilitates short-term retention in the rat. *Journal of Comparative and Physiological Psychology,* 1974, *86,* 161–171.

Roberts, W. A., & Grant, D. S. Interaction of sample and comparison stimuli in delayed matching to sample with the pigeon. *Journal of Experimental Psychology: Animal Behavior Processes,* 1978, *4,* 68–82.

Roberts, W. A., & Kraemer, P. J. Recognition memory for lists of visual stimuli in monkeys and humans. *Animal Learning and Behavior,* 1981, *9,* 587–594.

Roitblat, H. L. Codes and coding processes in pigeon short-term memory. *Animal Learning and Behavior,* 1980, *8,* 341–351.

Roitblat, H. L. The meaning of representation in animal memory. *The Behavioral and Brain Sciences,* 1982, *5,* in press.

Rosenfeld, S. A., & Van Hoesen, G. W. Face recognition in the rhesus monkey. *Neuropsychologia,* 1979, *17,* 503–509.

Ruggiero, F. T., & Flagg, S. F. Do animals have memory? In D. L. Medin, W. A. Roberts, and R. T. Davis (Eds.), *Processes of animal memory.* Hillsdale, NJ: Lawrence Erlbaum Associates, 1976.

Rumbaugh, D. (Ed.) *Language learning by a chimpanzee: The Lana project.* New York: Academic Press, 1977.

Sackett, G. P. Monkeys reared in isolation with pictures as visual input: Evidence for an innate releasing mechanism. *Science,* 1966, *154,* 1468–1473.

Sands, S. F., & Wright, A. A. Serial probe recognition performance by a rhesus monkey and a human with 10- and 12-item lists. *Journal of Experimental Psychology: Animal Behavior Processes,* 1980(a), *6(4),* 386–396.

Sands, S. F., & Wright, A. A. Primate memory: Retention of serial list items by a rhesus monkey. *Science,* 1980(b), *209,* 938–939.

Shepp, B. E. From perceived similarity to dimensional structure: A new hypothesis about perceptual development. In E. Rosch & B. B. Lloyd (Eds.), *Cognition and categorization.* Hillsdale, NJ: Lawrence Erlbaum Associates, 1978.

Shimp, C. P. Organization in memory and behavior. *Journal of the Experimental Analysis of Behavior,* 1976, *26,* 113–130.

Smith, L. B., & Kemler, D. G. Levels of experienced dimensionality in children and adults. *Cognitive Psychology,* 1978, *10,* 502–532.

Spear, N. E. Retrieval of memory in animals. *Psychological Review,* 1973, *80,* 163–194.

Spear, N. E. Retrieval of memories. In W. K. Estes (Ed.), *Handbook of learning and cognitive processes,* Vol. 4. Hillsdale, NJ: Lawrence Erlbaum Associates, 1976.

Spear, N. E. *The processing of memories: Forgetting and retention.* Hillsdale, NJ: Lawrence Erlbaum Associates, 1978.

Spear, N. E. Extending the domain of memory retrieval. In N. E. Spear and R. R. Miller (Eds.), *Information processing in animals: Memory mechanisms.* Hillsdale, NJ: Lawrence Erlbaum Associates, 1981.

Spear, N. E., & Miller, R. R. (Eds.) *Information processing in animals: Memory mechanisms.* Hillsdale, NJ: Lawrence Erlbaum Associates, 1981.

Spence, K. W. The nature of discrimination learning in animals. *Psychological Review,* 1936, *43,* 427–449.

Stollnitz, F. Forgetting of discrimination learning set by rhesus monkeys. Paper presented at Psychonomic Society meetings, San Antonio, 1970.

Stone, C. P. (Ed.) *Comparative psychology.* 3rd Edition. New York: Prentice-Hall, 1951.

Straub, R. O. *Serial learning and representation of a sequence in the pigeon.* PhD Thesis, Columbia University, 1979.

Straub, R. O., & Terrace, H. S. Generalization of serial learning in the pigeon. *Animal Learning and Behavior,* 1981, *9,* 454–468.

Straub, R. O., Seidenberg, M. S., Bever, T. G., & Terrace, H. S. Serial learning in the pigeon. *Journal of the Experimental Analysis of Behavior,* 1979, *32(2),* 137–148.

Terrace, H. S. *Nim.* New York: Knopf, 1979.

Terrace, H. S., Straub, R. O., Bever, T. G., & Seidenberg, M. S. Representation of a sequence by a pigeon. *Bulletin of the Psychonomic Society,* 1977, *10,* 269.

Terry, W. S., & Wagner, A. R. Short-term memory for 'surprising' vs. 'expected' USs in Pavlovian conditioning. *Journal of Experimental Psychology: Animal Behavior Processes,* 1975, *104,* 122–123.

Thompson, R. K. R., & Herman, L. M. Memory for lists of sounds by the bottlenosed dolphin: Convergence of memory processes with humans? *Science,* 1977, *195,* 501–503.

Tinklepaugh, O. L. An experimental study of representative factors in monkeys. *Journal of Comparative Psychology,* 1928, *8,* 197–236.

Tinklepaugh, O. L. Multiple delayed reaction with chimpanzees and monkeys. *Journal of Comparative Psychology,* 1932, *13,* 207–243.

Tulving, E. Episodic and semantic memory. In E. Tulving & W. Donaldson (Eds.), *Organization of memory.* New York: Academic Press, 1972.

Underwood, B. J. Attributes of memory. *Psychological Review,* 1969, *76,* 559–573.

Wagner, A. R. Expectancies and the priming of STM. In S. H. Hulse, H. Fowler, and W. K. Honig (Eds.), *Cognitive processes in animal behavior.* Hillsdale, NJ: Lawrence Erlbaum Associates, 1978.

Wagner, A. R. SOP: A model of automatic memory processing in animal behavior. In N. E. Spear & R. R. Miller (Eds.), *Information processing in animals: Memory mechanisms.* Hillsdale, NJ: Lawrence Erlbaum Associates, 1981.

Wagner, A. R., Rudy, J. W., & Whitlow, J. W. Rehearsal in animal conditioning. *Journal of Experimental Psychology,* 1973, *97,* 407–426.

Wickens, D. D. Encoding categories of words: An empirical approach to meaning. *Psychological Review,* 1970, *77,* 1–15.

Williams, B. A. Short-term retention of response outcome as a determinant of serial reversal learning. *Learning and Motivation,* 1976, *7,* 418–430.

Wilson, M. Identification, discrimination, and retention of visual stimuli, In A. M. Schrier & F. Stollnitz (Eds.), *Behavior of nonhuman primates,* Vol. 5. New York: Academic Press, 1974.

Worsham, R. W. Temporal discrimination factors in the delayed matching-to-sample task in monkeys. *Animal Learning and Behavior,* 1975, *3,* 93–97.

Zentall, T. R., & Hogan, D. E. Short-term proactive inhibition in the pigeon. *Learning and Motivation,* 1977, *8,* 367–386.

Zentall, T. R., Edwards, C. A., Moore, B. S., & Hogan, D. E. Identity: The basis for both matching and oddity learning in pigeons. *Journal of Experimental Psychology: Animal Behavior Processes,* 1981, *7,* 70–86.

Zoloth, S. R., Petersen, M. R., Beecher, M. D., Green, S., Marler, P., Moody, D. B., & Stebbins, W. Species-specific perceptual processing of vocal sounds by monkeys. *Science,* 1979, *204,* 870–873.

# Author Index

Numbers in *italics* refer to pages on which the complete references are listed.

## A

Abramowitz, R. L., 139, *150*
Adelman, H. M., 43, *48*
Allen, L. K., 105, *124*
Alloway, 180
Anderson, J. A., 133, *149*
Anderson, J. R., 175, *191*
Anderson, N. H., 14, *15*
Anderson, R. E., 133, *151*
Armstrong, G. D., 46, *49*, 60, *74*
Atkinson, R. C., 132, 134, *150*
Austin, P., *191*
Averill, J. R., 47, *48*

## B

Baddeley, A. D., 133, *150*
Balaz, M. A., 181, 191, *193*
Balda, R. P., 22, *31*, 189, *191*
Baril, L. L., 45, *49*
Barker, L. M., 45, *48*
Baum, W. M., 11, *15*
Beecher, M. D., 189, *196*
Bennett, R. W., 105, 121, *122*
Berryman, R., 56, *74*
Best, M. R., 45, *48*
Bever, T. G., 6, 8, 13, *15*, *16*, 28, *32*, 182, *196*

## C

Bierley, C. M., 107, *123*, 179, *193*
Birtwistle, J., 108, *123*
Bishop, T. M. M., 13, *15*
Bjork, R. A., 21, *31*, 138, 139, *150*
Blough, D. S., 24, 29, *31*, 56, *73*, 102, *122*
Blough, P. M., 24, 25, 29, 30, *31*
Bolles, R. C., 5, 6, *15*, 155, *171*, 174, 181, *191*, *193*
Born, D. G., 105, *124*
Bottjer, S. W., 55, *73*
Bower, G. M., 175, 181, *191*, *192*
Braggio, J. T., 116, *122*
Branch, M. N., 5, 6, *15*
Bregmeen, J. H., 176, *193*
Brewster, R. G., 17, *32*
Brodigan, D. L., 46, *48*, 60, *74*
Brogden, W. J., 43, *48*
Bryan, R. G., 156, *171*
Buchanan, J. P., 116, *122*
Bush, C. T., 181, *194*
Butterfield, E. R., 11, *16*

Cable, C., 187, *193*
Cacheiro, H., 181, 191, *193*
Campbell, B. A., 178, *192*
Capaldi, E. J., 175, 176, 180, *192*
Carley, J. L., 103, *123*, 141, *150*

197

# Subject Index